Sustaining the Earth

Sustaining the Earth

John Young

Harvard University Press
Cambridge, Massachusetts
1990

For Stephen, Philip, Susan, and Lindsay

Library of Congress Cataloging-in-Publication Data
Young, John, 1934–
Sustaining the earth / John Young.
        p.   cm.
Includes bibliographical references (p.   ).
Includes index.
ISBN 0-674-85820-4 (acid-free paper) : $19.95
1. Nature conservation—Political aspects.   2. Environmental
protection—Political aspects.   I. Title.
QH75.Y68   1990
304.2'8—dc20
90-4856
CIP

Contents

Acknowledgements

The idea of writing this book was suggested to me by Hugh Stretton longer ago than I like to remember. It was he who first aroused my interest in the politics of the environment, and eventually got me involved in the teaching programme of the Graduate Centre for Environmental Studies at the University of Adelaide. I give thanks to Hugh for that, and for his continued encouragement. I am also indebted to my other colleagues in the History Department, to which I have belonged for most of my academic life so far, for their toleration of my changing interests and enthusiasms, and for allowing me to accept a term of secondment to Environmental Studies for the past year. I am also grateful to my colleagues in the Centre, Pam Keeler, Ken Dyer, David Corbett and George Dubas, for the support they have given me throughout what has been a turbulent period for all of us.

My wife, Ruth, has made me get the thing finished after sharing the whole experience of searching in libraries, typing the first draft, listening to bits of it, pedalling a bicycle all over Britain and living for some time in a tent. Our thanks to Doug Carlston and Mary Crowley of Sausalito, Carroll Pursell and Angela Woolacott of the University of California at Santa Barbara, for their contacts and hospitality, and especially to Chuck and Alice Carlston who lent us their cabin in Maine in which to write. Laura Pines helped a lot, too, by coming over from her teepee by the pond one evening and lending us her tapes of Haydn's early symphonies.

I owe a great deal to the helpful criticism of Merrilyn Julian and many others who have read early drafts or parts of them; yet more to Marion Pearce, Silvana Flacco and Pam Keeler who put up with the resulting changes in their handiwork on the word-processor. Finally, my thanks to Iain Stevenson of Belhaven Press for his patience.

Introduction

The environment has suddenly been recognised by many of the world's leaders as an issue of central rather than peripheral importance. Many ordinary people have long suspected that, of all the problems which confront modern society, humanity's abuse of the environment is the one which holds the key to all the rest. It may be useful to have some kind of guide through the storms of controversy, the overfalls and counter-currents, the blind channels, shoals and reefs which lie between this suspicion and deciding what to do about it. So this book is something between a log, recounting such a passage, and, to conclude my metaphor, a book of pilotage.

A common starting point for such a quest is one or other of those challenging books which so effectively shook the complacency of those who believed in the inevitability of progress, and which first drew popular attention to the environmental problems of modern industrial society; books like Rachel Carson's *Silent Spring*, Edward Goldsmith's *Blueprint for Survival* and Paul Ehrlich's *The Population Bomb*. They argued strongly, from different viewpoints, that unless contemporary policies were drastically and rapidly altered the world faced a future of unprecedented calamity.

But thoughtful readers soon found that the answers these writers gave to the problems they revealed raised in turn new questions which greatly extended the potential scope of the enquiry. The new questions were about things like equity and poverty, about technology and energy, about aid and trade between rich and poor countries, and about the ways which people have had of thinking about them. They invoked powerful ideologies, strong emotions and deep-seated prejudices.

There is a rough parallel between the individual experience of finding a solution to one problem only to be confronted by new ones and the actual history of ideas about

the environment and society's relationship with it. By learning about the history it is easier to understand the ideas and
their implications and the way they grew from each other,
because they have developed, at different times and places,
and unevenly, but by means of a dialectic progression from
simple to more sophisticated.

So the first chapter of this book begins by recapitulating
that first realisation, back in the 1960s, when most of us
became aware for the first time that we lived in a finite
world which we had the power to destroy, not only quickly,
by blowing ourselves up, but more slowly, by allowing
existing trends to continue. It goes on to discuss some of the
arguments which followed that realisation, the strong feelings they aroused, and the revelation of the unexpected
complexity to which those arguments led. It tries to show
why the problem of what to do about it became not merely
technical but also economic, political and moral.

One useful definition of the relationship between a society
and its environment is culture, embracing as it does
religious beliefs and value systems as well as art, technology
and science. Chapter 2 therefore discusses some of the distinctive features of some societies which have retained a
sustainable relationship with nature. It does not argue that
we in the industrial world should return to such a state of
society, but that, in spite of the practical disadvantages of
'living as if nature mattered', pre-industrial society does
have lessons which might be usefully incorporated into the
philosophy of a sustainable society. This provides a perspective, in Chapter 3, from which to isolate historically both
aspects of industrial culture which have led to our present
environmental problems and those parts of it which provide
a basis for optimism about the future.

Chapter 4 discusses the role of science and its historical
relationship with society. It has been seen by many people
as the cause of several of the environmental problems of the
modern world. Some have resisted the suggestion of scientific or technological determinism, while others have
believed that if environmental problems were scientific in
nature then science could be expected to provide solutions.
The chapter discusses the sociological reasons for the failure
of this strategy so far.

Chapter 5 introduces the ideas of F.E. Schumacher and the

tradition of protest against the social and psychological effects of industrialisation which goes back to the Luddites. It considers the philosophical basis of the 'intermediate technology' movement, and its application and effectiveness in both rich and poor countries. The change in development policies since the optimism of the 1950s affords an institutional parallel which provides hope for the future. The environment, seen first as a technological and scientific problem, then as an economic and political one, has become a philosophical and ethical one.

Chapter 6 distinguishes between 'environmentalism', a reformist philosophy which maintains an essential distinction between the human species on the one hand, and 'nature' on the other, and a variety of more radical ideas which perceive humanity as inextricably linked with the rest of the biosphere. It discusses the relationship between these ideas and the philosophical diversity which has developed in recent times within the green movement in response to evidence of damage to the earth's atmosphere, climate change and the social crises of industrial society.

Chapters 7 and 8 discuss the implications of these changes for effective action. The first requirement is the recognition of common threads running through a very diverse range of single issue, political and religious movements, transcending cultural boundaries but drawing strength from a number of cultural traditions. The transition from ideas to politics is likely to be favoured by growing evidence that the kind of environmentally destructive development which has characterised the last thirty-five years will become impossible to continue in any case because of unforeseen social dislocation and the failure of present kinds of growth to cure it.

Opportunities for effective action will vary with time, place and individual circumstance. They require the diverse initiatives of people who are both intellectually and morally comfortable with their chosen course of action. They may agree that action is urgent but they will not necessarily agree on priorities. That should not prevent them from doing what they think is right. The process of trial and error which this book describes is likely to continue, and diversity of action is a necessary part of the collective experience of working towards a sustainable society.

Chapter one

Why we need a degraded environment

The writers who first shocked the world out of its complacency and drew widespread attention to the damage we were inflicting upon our environment now seem to have been over-pessimistic. Some of them expected Domesday to have come by now, but in spite of a prophecy of global famine in 1975 (Paddock and Paddock 1967), the relatively affluent reader of this book will have lived through it. Nineteen eighty-four did turn out to be somewhat as George Orwell predicted, but by the time it arrived, we had grown so used to things like 'doublethink' and 'newspeak' that we hardly noticed them. In rich countries we have not been directly affected by the famines which have devastated parts of Africa, and we have somehow adjusted to a growing incidence of cancer and other diseases linked with various forms of pollution. We have also grown used to a rising level of social violence and high unemployment. War in the Middle East, South East Asia, Africa, Latin America, and now China, is continuous and terrorism has become one of the accepted risks (except by insurance companies) of international travel. Plans have been made quite seriously for life after a nuclear war between the major powers, and as, during the 1970s, the price of oil fell and rates of unemployment began to stabilise at around 10 per cent, the 'eco-nuts' of the late 1960s began to take their place in the perspective of history. There are still plenty of politicians, economists and voters who believe that it was a mistake to have taken them so seriously (Simon and Kahn 1984).

However, if we look more closely at what has happened

since the mid-1960s, it looks not as if they were too anxious, but that they were anxious about the wrong things. The impending global catastrophe was seen at first as the simple result of technological excess, of man's inability to control the monster he had created. Scientific research, the basis of modern technology, of a polluting industry, of the exhaustion of resources and of uncontrolled growth at the expense of the environment, was seen as a kind of 'American roulette'. Most of the chambers of the revolver are crammed with twenty-dollar bills, but one of them has a bullet in it. If you are unlucky enough to find the bullet, you never get to spend the money you have accumulated while your luck lasted, and your good luck, most of the time, is totally negated by one unlucky event (Martin 1975). The so-called 'deaths' of Lake Erie and Lake Baikal were bullets, as was the evidence produced by Rachel Carson on the ecological effects of pesticides and herbicides in her book, *Silent Spring*, of 1962. There was no doubt, however, that there were plenty of twenty-dollar bills, and if only the bullets could be avoided by understanding how the mechanism worked, then everyone could share the proceeds of economic growth.

Dr Carson argued that the use of 'traditional' chemicals such as arsenic to control 'weeds' and 'pests' was harmful enough, but their effects were minor compared with those of the new synthetic hydrocarbons which were marketed aggressively and used irresponsibly. She drew attention to the action of the new substances which entered the soil and water systems of the world (and stayed there, as Australian beef producers learnt to their cost in the 1980s). She argued that the methods of 'applied entomology' were, in the first place, counter-productive because the insect world was successfully responding through rapid micro-evolution to produce insects increasingly resistant to poisons which had once been fatal. Each new and stronger poison created to meet the challenge increased the risk and the incidence of man poisoning birds, animals and fish, which he had no intention of harming. Man thus poisoned his own food and ultimately, himself, as shown by the greatly increased incidence of cancer among the American population and the increasing youth of cancer victims as the twentieth century progressed.

But Rachel Carson was perhaps over-concerned about the possible reaction of her scientific colleagues and did not wish to seem 'political'. She did hold chemical companies and their salesmen responsible to some extent, and she also blamed farmers and local governments, but she drew no general political conclusions. Her remedy was biological control, though sometimes with the aid of a little judicious and selective chemistry to render insect populations infertile. More important, for its impact on her public readership, was her questioning of the value-system of a society which assumed the right as well as the ability of humanity to manipulate nature and to escape the costs of its *hubris*.

None of the bullets proved to be fatal, but they made thoughtful scientists pause in their stride and, for the first time, the complacency of a generation educated to take science on trust was shaken. It was still believed that the problems of which the 'silent spring' was a symptom were problems of the misuse of certain aspects of science. Some scientists and many more laymen were later to respond in the same way to the nuclear accidents at Three Mile Island and Chernobyl. Early mistakes, it was assumed, could be rectified quite simply when the problems were fully understood, within the appropriate scientific framework. It was assumed also that scientists, who were the experts in the field, would take the lead in providing whatever remedies or safety measures might be needed.

This alone gave grounds for optimism because it represented a significant change of heart on the part of the scientific community. It was fashionable in the first few decades of the twentieth century for scientists to disclaim moral responsibility for the results of their discoveries, but the new generation of veterans who returned from the Second World War, and those who entered academic life with them, did not shrink from the implications of omnipotence. Many of them were ready to add to their agenda not only the search for scientific solutions to particular environmental problems, but also the social and political problems which were now seen as at least incidental to them.

The first environmental crusaders of modern times were therefore people like Paul Ehrlich, a professor of biology, author of *The Population Bomb* (1968) and the group of ecologists, supported by thirty-seven eminent chemists, zoologists, medical scientists, microbiologists, botanists, together with an archaeologist and an economist, who compiled the *Blueprint for Survival* (Goldsmith 1972).

Both books were designed to shock and were written in the characteristically academic belief that the main obstacle to intelligent action is ignorance rather than conflicts of interest: once the problem was explained with sufficient clarity and authority, then voters and governments alike would act on the advice which scientists were able to give them.

Ehrlich's book begins with some frightening statistics leading to a fantasy 900 years hence when the world's population of 60 million billion people will need to be housed in a continuous 2,000-storey building covering the entire planet. The *reductio ad absurdum* device is then set aside and the current manifestations of an environmental crisis are listed: industrial pollution, the extinction of species, the loss of wilderness, the inexorable spread of 'urban blight' creating a breeding ground for youthful disaffection, crime and violence. The causal chain 'is easily followed to its source . . . too many people'. Then follow a series of alternative scenarios. The first two end in nuclear war after a series of setbacks in American foreign policy combined with domestic environmental catastrophe. The third has as a major premise the death of one-fifth of the world's population by starvation and the collapse of effective government throughout Asia and Africa. The affluent nations, Japan, Australia and the European Community (EC) then group under an enlightened American leadership to control population growth, introduce controlled agricultural and industrial development and eventually stabilise world population at 1.5 billion in 2100.

The argument that these are the likely alternatives is then reinforced. 'Family planning' as presently preached and practised is the mere bailing of a sinking ship with a thimble because it is just that — planning to have families instead of planning not to have them. Potential food supplies from marginal lands or the sea, or from mineral sources, are dismissed as utopian, though some hope is placed in the 'miracle grains' of the 'green revolution' which have achieved high yields as a result of creative plant breeding. The best that can be hoped for, however, is a brief delay of inevitable global catastrophe unless urgent action to halt population growth is taken immediately.

The logic of the discussion leads to the conclusion that population *control*, by governments, not just voluntary planning, is our only hope. The first step is to control the population of the United States by way of example to the less enlightened. The best method, theoretically, might be to introduce temporary sterilants to water supplies, but Ehrlich admitted that there were both technical and political problems. However, there was much that a new and all-powerful Department of Population and Environment (DPE) might do on all fronts to achieve a cleaner North America. Sex education which played down reproduction, taxation of large families instead of tax relief for parents, prizes for sterilisation and industrial pollution policy based on the 'polluter pays' principle were the recommended initiatives.

When it comes to the 'underdeveloped' countries, Ehrlich supports the strategy suggested by Paddock and Paddock the previous year. America has the power and therefore the responsibility to introduce an 'aid' policy based on the 'triage' concept borrowed from military medicine. In the emergency of battle, resources must be used to the best possible advantage. Patients are therefore divided into the 'walking wounded', who will eventually recover without medical assistance and so do not constitute a priority, the 'living dead', whose case is so bad that they will die anyway, and on whom resources would be wasted, and those who can survive if they receive prompt and competent treatment, but who will die without it. Applying the analogy with ruthless logic to a number of needy countries, Ehrlich does not, as Paddock and Paddock do, bring the question of political allegiance to the United States into it: aid sent to India is regarded as wasted, Libya is judged to be a 'walking wounded' case in which oil revenues will buy the necessary supplies in any case, and West Pakistan is a suitable candidate for assistance.

In such lucky countries, transistorised TV sets should be distributed to 'educate' the masses. Migration should be rigorously controlled to prevent the 'swamping' of areas selected for development programmes, and effective population control becomes the overriding condition of continuing aid.

Ehrlich was preaching to a considerable number of the already converted. Typical of those ready to agree with him were the members of the Sierra Club of California, a group of wealthy but concerned conservationists who were quick to catch the underlying sense of distaste for un-American humanity which is revealed on the first page of Ehrlich's work. He recounts a nightmare taxi drive through the back streets of Delhi in 100-degree heat into the slums, which

seemed alive with people. People eating, people washing, people sleeping. People visiting, arguing and screaming. People thrusting their hands through the taxi window, begging. People defecating and urinating. People clinging to buses. People herding animals. People People People. . . Would we ever get to our hotel? (Ehrlich 1968, p. 15).

In England the already converted were similarly conservative, but the emphasis of their conservatism was different. They were concerned less than American conservationists about the psychological or genetic values of wilderness but were more concerned about the amenities of the cultivated British countryside, which seemed threatened by the demands of industry and the vulgarities of excessive numbers of city trippers. Just as *The Population Bomb* placed a parochial anxiety about Californian wilderness, linked with fears of urban ghettos, in a global context, so the *Blueprint for Survival* broadened the concept of conservation of the English countryside to that of a concern for sound environmental management, but pointed to a radical rather than a conservative solution.

It begins with a stern warning about 'the extreme gravity of the global situation today', and forecasts the impending breakdown of society and 'the irreversible disruption of the life-support systems on this planet possibly by the end of this century' if current trends are allowed to persist. It aimed, therefore, to found a 'movement for survival' based on a new philosophy of life in which man would learn to live as part of nature rather than as its antagonist.

Like Ehrlich, Goldsmith found the root of the problem in too many people. Indefinite growth confronted with finite resources must lead to disaster. If resources did prove more extensive than expected, that would lead to graver problems of waste disposal and heat dispersal. A new concern was the disruption of traditional societies by their increasing inclusion within the Western economy, which resulted in the heightening of aspirations which must, for the health of the planet, remain unsatisfied. Perceiving the political difficulties of getting any government to support a 'no growth' policy, the *Blueprint* argued that governments would only attempt it when it could be shown to be possible without causing unemployment. This would only be possible 'within the framework of a fully integrated plan'. Such a plan was then put forward.

The four conditions for a stable society were stated as minimal ecological disruption, the maximum conservation of materials and energy, zero population growth and the 'social conditions' in which the first three conditions could be enjoyed.

Forgetting for the moment that the creation of such conditions meant government action, of a kind which in Britain, not to mention other countries, was not expected while unemployment remained at a high level, Goldsmith called for a series of measures which might cause even the most determined of dictators to hesitate. The first steps were a growth freeze achieved by a tax on the use of raw materials, and an amortisation tax, graduated from a minimum rate on products designed to last 100 years or more to a maximum rate on plastic disposables. Further proposals were a power tax, the substitution of natural and renewable materials for plastics and non-renewable ones, and legislation to protect wilderness while allowing wilderness areas occupied by hunter-gatherers and 'primitive agriculturalists' to remain in the undisturbed custody of their indigenous inhabitants.

These reforms were intended to encourage employment-intensive industries, craftsmanship and creativity and to solve waste-disposal problems. They would also reduce pollution and tend to the preservation of the world's ecosystems. Direct legislation would be assisted by the universal application of the 'polluter pays' principle in taxation and the abandonment of the gross national product (GNP) concept as a measure of well-being. It was to be replaced by a system of 'social accounting' in which costs, such as loss of amenities, should be subtracted from production, such as the building of motorways.

The precondition for all of this was seen as zero population growth. Britain should go one better by aiming to reduce population to 30 million in order to be self-supporting and free of the necessity to import food from other countries. The target should be the replacement family throughout the world by the end of the century, to be achieved by advertising, free contraceptives, abortions on demand, and, sensing perhaps that there might be some rather intractable problems in this area, investment in 'socio-economic research' to *make* population control acceptable. A massive education programme would start immediately with the goal of community acceptance of the philosophy of a sustainable society by the time that everyone under fifty years old had been thus re-educated. All were ultimately to be convinced that 'there is no valid distinction between the laws of God and Nature and Man must live by them no less than any other creatures'.

The proposal for which the *Blueprint* was to become most memorable was decentralisation. Utopia was to consist of small village-type settlements of about 500 people, forming part of larger communities of around 50,000 in regions of half a million or so. Decentralisation would facilitate a reduction in capital costs through community self-sufficiency in such things as sewerage and water supplies. No new roads would be built. The money saved would go to subsidise public transport, especially railways, and to restore canals.

Instead of the social and ecological disaster represented by the modern megalopolis, 'human-scale communities' would create a new awareness of social responsibility and co-operative opportunity. Community loyalties, on the one hand, and global awareness, on the other, would leave little room for nationalism, that 'dangerous and sterile compromise'. Ominously, some critics thought, they would facilitate the kind of social pressure and moral coercion which radical reform might require. Others could see little wrong with the anonymity, the occasional and necessary loneliness even, complementary to that of wilderness, in the megalopolis, nor with the levels of excellence in art, drama and music, nor with libraries, universities and cathedrals, the intellectual stimulation and excitement of public debate which only a large city can support.

There were, in addition to the thirty-seven eminent

scientific supporters, a further 180 scientists, some equally eminent, who supported the ideas of the *Blueprint* in general but felt unable to subscribe to them completely, 'because it contained scientifically questionable statements of fact and highly debatable short term and long term policy statements' (*The Times*, 25 January 1972).

This was perhaps the first indication of a public recognition that the 'environmental crisis' was not susceptible to solution by the application of a single formula or even by a combination of specific remedies, such as the *Blueprint* had recommended. It was more than a combination of its constituent parts — pollution, resource depletion, famine and so on — and would require as much imagination as analysis to comprehend it, let alone do much about it. With the publication of Barry Commoner's *The Closing Circle* (1972) it began to be apparent that there was a basic difference between the 'problem of environment' and other very complex problems such as splitting the atom or getting a man onto the moon, and that, in ignoring this difference, scientists like those who wrote the *Blueprint* were making what philosophers call a 'category mistake'.

Commoner's book was typical of the end of what might be called the 'first round' of environmental awareness: much of his causal analysis was parallel to that of Carson, Ehrlich and Goldsmith, but he introduced a new idea, contained in two central chapters which he called 'The Technological Flaw' and 'The Social Issues'. The first argues that it is not growth or population as such which causes environmental damage, but the things which people do and, specifically, the new technologies which they have employed since the Second World War, which are mainly responsible. The second attempts to discover why they have been so damaging, and concludes that the very success of technological achievement — detergents in getting clothes 'whiter than white', nuclear bombs in causing maximum damage, car engines in producing more power — have been the predictable result of the scientific method.

It seems clear in retrospect that the huge advance of science and the increase in knowledge in the middle decades of the twentieth century created great pressure to isolate problems and specialise skills, and many of the spectacular advances came from doing that. There was some consequent

tendency for the specialised branches to attract most attention, and perhaps most talent. There continued to be
generalists of many kinds — ecologists, soil biologists, and
others — though their progress was less spectacular in terms
of new technical and scientific breakthroughs.

But while science was making its most spectacular
achievements by isolating problems and processes, environmental prudence demanded an opposite strategy: an obser-
vant generalism, an insistence on relating scientific
problems to their ecological context, and scientific solutions
to their ecological consequences. The social issue, which lay
behind the environmental crisis, was the failure of both
public and politicians to differentiate between two kinds of
decision: technical decisions about means, and moral decisions about ends. It was sensible to call in scientific expertise once a decision to do something had been made in the
full knowledge of its possible consequences. Making such
decisions, whether about developing local industries which
might both create jobs and pollute lakes, or nerve gases,
which might defend the nation but kill local livestock when
improperly stored, involved making value judgements about
what was wanted, and in making that kind of judgement
scientists had no monopoly of competence.

Commoner goes on to show how, in one area after
another, American society had fallen foul of this confusion,
but concludes by disclaiming the optimism of his transatlan-
tic colleagues:

In our progress-minded society, anyone who presumes to explain
a serious problem is expected to offer to solve it as well. But none
of us, singly or sitting in Committee, can possibly blueprint a
specific 'plan' for resolving the environmental crisis. To pretend
otherwise is only to evade the real meaning of the environmental
crisis: that the world is being carried to the brink of ecological
disaster not by a singular fault, which some clever scheme can
correct, but by the phalanx of powerful economic, political and
social forces that constitute the march of history. Anyone who
proposes to cure the environmental crisis undertakes thereby to
change the course of history.

But this is a competence reserved to history itself, for sweeping
social change can be designed only in the workshop of rational,
informed, collective social action. That we must act is now clear.
The question which we face is how (Commoner, 1972, p. 300).

Commoner's book was construed by some as an attack on science itself, and it evoked a strong response from science-minded optimists. Professor Burhop, for example, at the World Conference on 'Science and Society' of 1971, expressed a widely held view when he said that 'the problems science has created need, not less science, but more science, for their solution' (Clarke 1975, p. 120). Others maintained that Commoner's arguments about the environmental effects of particular lines of research which had led to high explosives, or a polluting plastics industry, could be made to seem dangerously close to anti-intellectualism. Was it not more in keeping with the principles of academic freedom to assert that if cars caused pollution, then anti-pollution devices should be invented?

John Maddox was a theoretical physicist who had become influential at an academic level as a director of the Nuffield Foundation, as Affiliate of the Rockefeller Institute of New York, and also at a popular level, as a broadcaster and as Science Correspondent to the *Guardian*. His book, *The Domesday Syndrome*, published in 1972, was thus authoritative and reassuring to those who wanted to dismiss Ehrlich, Carson and Commoner as erudite but misguided extremists.

He argued that the most obvious solution to the problem of famine was to make fuller use of 'unproductive' land: 'In many parts of the world, but especially in the tropics, vast areas of land are given over to unproductive rain forests' (Maddox 1972, p. 26). Beyond that, the 'Green Revolution' had the answer. New strains of high-yielding rice could feed the world and at the same time raise living standards and reduce international tension by introducing incentives for capital accumulation. Aid programmes could be designed to undermine the social conditions which bred insurgency and communism. Energy had now surfaced as a major environmental problem next to pollution, but it was dismissed initially as a matter of ignorance on the part of non-scientists, in view of the inexhaustible possibilities of nuclear power and hydrogen as sources of energy. The 'Domesday Men' were guilty, it seemed, not of misunderstanding the complexity of the problem or of discounting the political implications of their remedies, but of simple exaggeration and lack of faith in the ability of science to cure what it may have caused. There were, it was acknowledged, social, economic and political 'aspects' to what was an undoubtedly scientific problem, and it was conceded, as a matter of academic good manners, that there was certainly a subsidiary contribution to be made by the social sciences. The overriding question, wrote Maddox (1972, p. 69) 'is not whether cleaner air can be provided, but how much tax-payers, and in particular, the operators of motor vehicles and factories are prepared to pay for that amenity'.

Herman Kahn was a founding father of the Hudson Insti-tute of New York, established in 1961 to engage in policy analysis, especially on issues related to American national security. However, it became, in Kahn's words, 'more and more engaged in the infant but growing field of future studies'. It was largely funded by grants from US govern-ment agencies and multinational corporations. In 1967, Kahn published, with A.J. Wiener, *The Year 2000*, which was concerned with the relatively long-term prospects of the human species. This was followed in 1972 by Kahn and Bruce-Biggs's *Things to Come: Thinking about the Seventies and Eighties*, with a 'tactical' rather than a 'strategic' emphasis, and published through normal commercial chan-nels. It was designed to acquaint the layman with the methodology and the credibility of the infant science of *futurology*.

The authors saw little to worry about in the views of those who were beginning now to sense that the problems of the environment were merely symptomatic of a much wider network of interrelated problems such as multiple inequality between rich and poor nations, and between rich and poor in both of them; the monopoly of power by big business, big government and big bureaucracy; and that the key to environmental reform might be social change.

'No' says Kahn. 'In 1985 the people of the world will be more culturally similar than they have been at any time in the history of mankind', and, 'to a remarkable degree, this 'global metropolis' will be Americanised' (Kahn and Bruce-Biggs 1972, p. 19). At this stage the success of revolutionary movements in Vietnam and Nicaragua were not foreseen, nor was the strength and anti-Americanism of the Islamic revival, but there was much to support the assumption of expanding American cultural imperialism in the wake of an expanding global economy. The middle-class faddish critique of the work ethic was a problem closer to home, as were increasing violence, insurgency, criminality, unemployment and drug addiction. The most likely result of an American defeat in Vietnam was seen in 1972 as national indignation and a witch hunt for treasonable liberals for undermining the resolve to win. There is no inkling of any interruption to the continuous growth of the American GNP, of the oil crisis, which took place a year after the book was published, or of Watergate, which took the sting out of any right-wing backlash which might otherwise have followed Vietnam. Ultimately, for Kahn, power is the key to whatever reforms may be needed in the interests of environmental management and the social problems incidental to it. Technology gives power, and for those who have it, continued optimism is legitimate: 'Some forms of technology are needed just to cancel each other out ... if there are new techniques for insurgency, criminality or ordinary violence, we need new techniques for counter insurgency or imposition of order' (Kahn and Bruce-Biggs 1972, pp. 214–15).

Being wrong about the oil crisis, the green revolution, the integrity of the presidency and the consequences of Vietnam might lead us to question the thesis as a whole, but the culturally specific nature of the 'hard-nosed realist' reaction soon led to a further widening of the terms of reference of the debate and a questioning of the goals and assumptions which not only Kahn, but also Paul and Anne Ehrlich and their contemporaries had taken for granted; should not the assumption that the lifestyle of the wealthiest nation on earth was a model for emulation by the rest of the world be challenged in any case, some asked, as a prerequisite for environmental reform?

'Third round' theorists about the environment include a number of writers who, though having individual priorities and perspectives, probably would not object to being described as adherents of the humanist left (Clarke 1975, Stretton 1976; Sandbach 1980; Schnaiberg 1980). They were just as critical of the 'old left' (or what was called the 'new left' in the 1960s) as they were of the more recent 'hard-nosed right', not so much for being politically disagreeable as for missing the point. To speak of an 'ecological crisis', for example, is to confine the debate, they would say, to an area of ineffectiveness. To quote Clarke (1975, p. 62):

In many ways the ecological crisis was, and is, a cover up.
Throughout the 1960s the New Left, particularly in the United
States, sought to expose it as a liberal plot to divert attention from
the real problems, those of social injustice, war and the evils of
capitalism [Stretton would add the evils of state socialism as well].
The truth is, of course, that the ecological crisis was not a diversion from these things, but a result of them.

These writers, none of whom are scientists, agreed that the
dream of a science-produced cornucopia of human happiness
had been replaced by something of a nightmare, not because
the accomplishments of the individual sciences have been
flawed, but because of the political circumstances in which
they have been achieved. Their consensus about the
relationship between environmental problems and the
political context in which they arise can be summarised as
follows.

It was in the period of rapidly rising production after the
Second World War that the environmental crisis first made
itself felt in the form of globally dangerous degrees of pollution. The trouble was that with basic welfare and full
employment in the fluid, expanding Western economies,
even the poorest were better off than they had ever been
before. Inequalities no longer had to be permanent it
seemed, and if pollution was the price which had to be paid,
it was going to have to get a lot worse before majorities took
it seriously. It was getting worse, but in ways which were
only apparent to those who made it their business to
measure such things rather than to analyse the causes of the
processes they were measuring. Meanwhile, it was what
seemed to be the beneficial end-products of industrial
processes rather than their problematical side-effects which
gained the attention of the majority and won its approval.

The technologies which arose from the scientific
discoveries of the inter-war period had been utilised as
devices under pressure of war itself, and were now available
for application to peaceful production, especially in the
chemical industries, synthetics, detergents, and construction
materials like glass and aluminium which needed large
amounts of energy for their production resulting in corres-
ponding amounts of pollution. Barry Commoner (1972,

p. 258) had already pointed out, that the reason for this movement was that the profit of the more polluting industries was higher than in the less polluting ones. Novelty encouraged price fixing and required a heavily polluting energy- and resource-consuming advertising and packaging industry, to keep up consumption.

Other humanists, especially concerned with the so-called 'Third World', pointed out that inequality between rich and poor nations, and between rich and poor in both of them, was the context in which the choices which determined the directions of industrial growth were made. The result was the negation of the good effects which science might otherwise have had. The food crisis, for example, which might have been eliminated in much of Asia and Africa if science had been applied to the improvement of local and subsistence production, became worse. This deterioration was related not only to population growth, but also to the use of much of the best arable land and scarce water resources for the production of raw materials which were exported to the industrial world to pay for imported goods consumed by the indigenous elites. Even when land and water resources were used for food production, that production increasingly became industrialised, favouring large producers who were able to import machinery, fertilisers and pesticides from the industrialised world. Often, but not always, this increased output, at least temporarily, but it often increased inequality as well, in rural areas, and made increasingly scarce industrial employment for the poor a novel prerequisite for getting enough to eat.

Thus in much, though not all, of the 'Third World,' the monopoly of wealth by local elites and their demand for the consumer goods of industrial society has produced structural underdevelopment and political instability.

It became clear that in rich countries, inequalities were once more increasing, especially in the decade after 1975 in which the environment came increasingly under discussion. A high rate of unemployment had now become an accepted feature of capitalist societies, and the proportion of national budgets expended as dole payments increased in absolute terms as more people became unemployed. Some economists argued that this was the inescapable and therefore acceptable cost of the rationalisation of industry in the context of new technology. Others thought it followed from workers pricing their labour too high; these theorists called it all 'voluntary unemployment' and many blamed it on the existence of a too easily available dole. Affluent people, taking Ehrlich's advice, began to have fewer children, but the independence this gave them and the money they saved enabled them to buy several cars per family, while the families of the unemployed poor depended once more on public transport systems which were being wound down as a result of two decades of a losing battle with universal car ownership.

So what started out in the 1960s as a feeling of emergency, calling for immediate and drastic government action in one or two industrialised countries, now looked as if it might have to go on the 'back burner' for a while, in view of its unforeseen complexity. The environment was not merely a technical but also a political problem, and it concerned not just national politics but also international politics and economic relations, fields in which it was notoriously difficult for mere pressure groups to have much influence. Environmentalists had rather hoped they could leave their political allegiances and private interests on one side as they petitioned for airports to be located somewhere other than where they spent their weekends fishing, or for quarries within earshot to be closed, but it was not to be.

While multiple inequalities remain the overriding concern of party politics, (right-wing parties being anxious to maintain them and left-wing ones to reduce them) legislation aimed at improving the environment is often able to gain considerable support from a concerned middle class, but not from majorities, especially if it can be presented, and it always can, as involving either the loss of more jobs or higher costs to the consumer. Majorities in rich countries are not likely to favour radical reform of aid policies to poor countries while they feel that their own economic problems are not being addressed effectively. 'Charity begins at home' is a useful slogan when there is little chance of it beginning anywhere.

Majorities in both rich and poor countries are thus forced, in their own immediate self-interest, to support a pattern of development which leads to an increasingly narrow range of choices, all with dangerous long-term implications.

Humanists point out that the theory of consumer sovereignty, which might have provided grounds for optimism, has been invalidated by the development of monopoly capitalist production. As Galbraith (1971) points out, 'The risks of competition are avoided by enterprises able to avoid them, while maintaining a public ideology of the virtues of the free enterprise system'. Consumers as such are thus exonerated from the blame for environmental degradation resulting from such things as overpackaging and overproduction because of the degree to which consumer demand is now susceptible to management:

Few producers of consumer goods would care to leave the purchases of their products to the spontaneous and hence unmanaged responses of the public. Nor, on reflection, would they have much confidence in the reliability of their labour force in the absence of pressures to purchase the next car or meet the next payment on the last (Galbraith 1971, p. 267).

Instead of production being controlled by consumers, producer sovereignty, in the rich countries, is closer to the reality. The maintenance of what amounts to a system of producer control through the manipulation of consumption is enhanced by the advantages enjoyed by multinational corporations. They often control supplies of raw materials, and the unprecedented profit volumes of the 1950s and 1960s permitted the largest corporations to become self-financing. Costs were therefore not minimised, since the question asked was not 'what is a reasonable profit margin?' but 'what price will the market stand?' Expansion ensured that the answer was usually 'more' as choice became more limited and it remained possible to pass costs further down the chain of consumption, and onto future generations.

In the same period, and up to the present, the role of government in the rich countries has been to maintain a balance between economic development of a kind which engenders social inequality on the one hand, and social peace on the other hand, a peace which is threatened whenever a slowing down of economic development draws attention to its side-effects. Development results in increasing inequality because of the part played in it by big government, big business and big bureaucracy. If they do not favour capital intensive forms of growth, as opposed to decentralised, diverse and labour-intensive forms of growth, they are judged to be inefficient and will fail to justify their programmes to politicians who need to be able to show 'runs on the board' to an electorate trained by big media to value such yardsticks of well-being as GNP and balance of payments.

Majorities in rich countries find that, in spite of the figures, they seem less well off than they used to be. Average families in rich countries in the late 1980s find that they need two wages, when one is difficult enough to find, in order to maintain the level of consumption which they have been insistently taught to expect. They eat meat less often than they used to in the 1960s, yet there seems to be no lack of funds for such purposes as America's Cup yacht racing. It has only been possible to prevent majorities becoming aware of conflicts of interest when growth has been sufficient for the relatively small gains which do filter down to the poor to result in an absolute rise in living standards sufficient to offset the resentment caused by increasing aspirations in a context of increasing social inequality. It is a delicate balance at the best of times, and one which can easily be upset by the intrusion of additional factors such as racial or religious antagonism. Treading the tightrope successfully means supporting the expansion of monopoly capital and adopting policies aimed at distributing just enough social surplus to the poor to avoid too much social turbulence or the achievement of office by radical politicians. Leaders like Bob Hawke of Australia, Britain's Mrs Thatcher and President Mitterrand of France succeed with this feat very well in times of economic good fortune, based on high prices for minerals or farm produce, North Sea oil, or full overseas order books for the arms trade. Without such props, they know that they must expect trouble.

Avoiding trouble in rich countries thus has, as its corollary, the maintenance of present inequalities and continued environmental degradation. As the maintenance of growth on both sides of the Iron Curtain depends in turn on control of raw materials found increasingly on the periphery of the world economy, it depends also on the industrialised nations extending their control of the economies of countries in Asia and the Pacific, Africa and Latin America. The 'spheres of influence' and 'men on the spot' which heralded the crude imperialism of the nineteenth century are replaced now by 'fishing agreements' and 'freedom-fighters', often reinforcing or increasing inequalities in poor countries as well.

This is not to deny that development has brought many benefits, even when investment, trade and aid have had political strings attached. Individual opportunities have been greatly enlarged by improved means of transport, modern communications, medical services and access to Western education, but the side-effects of development have included the replacement of old kinds of inequality with new kinds, which have become institutionalised as complicated administration has demanded specialised bureaucrats who have required air-conditioned offices in which to work and imported cars in which to be driven about.

Meanwhile, the desire for cash to purchase imported goods, from transistor radios to exocets, means that in most poor tropical countries now, agricultural land is used more intensively; fallow periods are shortened, less nutritious crops like cassava become popular because they grow easily with less labour on poorer soils, as compared with yams or taro, for example, which are more nutritious. The dietary gap is filled, usually inadequately, by imports like factory-made biscuits, with sliced bread and tinned fish, creating, in due course, a demand for a high-technology medical industry to combat the ailments of 'affluence'.

Many forms of 'aid' clearly make these problems worse rather than better, especially when they take the form of military aid, or are used largely to pay for the salaries of expatriates in the service of foreign firms or governments, which are spent largely in the metropolitan countries.

Fiji is a fairly typical poor country — not as poor as some African countries, but poorer than Malaysia or the Philippines. In 1969 Ratu Mara, the man who was to become prime minister at independence and has now accepted the job again as Fiji pulls back from the brink of military dictatorship, became acutely aware of the uneven distribution of the wealth produced by development within his country, and gave an ominous warning to the expatriate community of Suva that if the whole town were to be burned to the ground, the Fijian people would lose nothing but the record of their debts.

On a world scale, the same can be said of the consequences of the destruction of New York, London, Paris and
Tokyo, for two-thirds of the world's population. There is
evidently a correlation between the ability of a society to
ensure adequate food and shelter for all its members,
including the poorest, and the degree of equality within that
society. Regardless of national wealth, relatively egalitarian
countries like Sweden have very few beggars. Unequal
societies, whether rich, like Britain, or poor, like Bangladesh
and Ethiopia, have many beggars. There is a similar link
between good global management and equality between
nations. Amazonian rainforest is not, after all, destroyed to
feed poor Brazilians with grain but to keep rich North
Americans in beef steak. The obvious remedy, in view of the
problems inherent in the development of poor nations, is to
decrease the relative wealth of rich nations. But European,
Australian, American and Japanese majorities are not likely
to support such a programme while their own societies grow
increasingly unequal and the numbers of their own poor
increase. We must, therefore, accelerate the degradation of
the environment to make the world safe for inequality.

Suggestions for further reading

Rachel Carson's classic *Silent Spring* is still as good a starting point as it ever was, together with her earlier *The Sea
Around Us* (1956) and *The Sea* (1964). Paul Ehrlich followed
his popular work, *The Population Bomb*, with a great deal of
detailed scholarly work including (with Anne Ehrlich and
J.P. Holdren) *Human Ecology, Problems and Solutions*
(1973) and *Eco Science: Population, Resources, Environment*
(1977) while the essential message of *Blueprint for Survival*
has been refined and exemplified throughout the period
since 1972 in the pages of *The Ecologist* which Goldsmith
edits. His recent book, *The Great U Turn* (1988b) is a collection of essays on the problems of industrial society and the
options which now face it.

Barry Commoner went on to develop his arguments
further in *The Poverty of Power: Energy and the Economic
Crisis* (1976) and *The Politics of Energy* (1979). Alan
Schnaiberg's book *From Surplus to Scarcity* (1980) is

textbook-like in its erudition and was also a landmark in the humanist literature. It remains useful, not only for its broad approach, but as a work of reference, and its bibliographies are detailed and excellent. Perhaps the most important book of this period, though, and central to the argument of this chapter, is Hugh Stretton's *Capitalism, Socialism and the Environment* (1976). The development of his ideas can be traced from *The Political Sciences* (1969), *Ideas for Australian Cities* (1970) to his more specialised *Urban Planning in Rich and Poor Countries* (1978). All of these, and many of the works republished in his prize-winning *Political Essays* (1987) deal with the issues of social equity which underlie the environmental crisis.

International inequality is the subject of a vast literature but most pertinent to the issues raised here are the works by Myrdal and Frank which are discussed below, p. 110, Paul Harrison's *Inside the Third World: The Anatomy of Poverty* (1987) and Malcolm Caldwell's *The Wealth of Some Nations* (1977).

To summarise the discussion so far: by the middle of the 1970s most people who were interested in the problems of the environment agreed that they had not arisen simply as a result of technological excess, or even of growth in population, but because of the technological circumstances of population growth and the political circumstances in which technology had developed. There were two kinds of response to this consensus; the first pragmatic, and the second humanist.

Pragmatists and humanists could agree at least on some things: that the problems were beyond the scope of simple single-discipline scientific remedies or 'bandaids'. But the pragmatists advocated the extension of the scientific method, from dealing with the physical universe, into the 'softer' areas of politics, social engineering and security, so that the ultimate benefits of growth and affluence could one day be enjoyed by everybody in a well-managed environment.

Humanists objected that this brave new cure might be worse than the disease and were inclined to argue that the blessings of growth regardless were not worth the inequality and oppression which might have to be the corollary of growth. This led many to conclude that comprehensive social change in the direction of greater equality was a prerequisite of environmental reform. This kind of radical reform would be difficult because, as long as growth continued, majorities found they could live with inequality.

One way out of this spiralling dilemma was the fundamentalist one, hinted at, but not explained, in Blueprint for Survival. This was the twentieth-century version of that recurrent ingredient of radical ideology in Western society, the myth of the noble savage, or a return, in the shadow of domesday, to the Garden of Eden. Some reformers, therefore, cultivated the belief that economic growth was a cultural aberration and that the tribal societies of hunter-gatherers and subsistence agriculturalists not only lived in harmony with their environments but also did not suffer from psychosomatic or stress-related diseases, or from social delinquency. They did not, therefore, require the costly social and medical services of modern society.

It was left mainly to young people to experiment in a wide variety of 'back to nature' movements, singly, in couples and in communes, in cities, countryside and wilderness, but most notably in the wildest or most beautiful parts of the wealthiest countries such as Wales in the United Kingdom, California and Maine in the United States and northern New South Wales in Australia. In some cases there was little time between the departure or extermination of the last traumatised remnants of an indigenous culture and the arrival of refugees from the culture which had displaced them.

Many of the experimental communities changed and grew and still survive. Some dissolved and some of the more bizarre were suppressed. Though characterised by great diversity, they all shared a belief that the morality on which industrial society is based is unsound. It was, therefore, to be expected that the reformers should look for inspiration to the societies which industrial society had sometimes quite recently displaced, and to the relationship which had been established between those societies and their environments. The rhetoric of North American Indians achieved a high profile in the literature of what was in origin and inspiration a North American movement, but some eco-activists sampled the ideas of a number of non-Western societies.

The trouble was that real experimental isolation was almost impossible to achieve and in times of crisis and often in times of mere difficulty, industrial society provided a safety net which critics were able to see as invalidating the experiment. One alternative to joining a commune is to read the next chapter (joining a commune might still be a good idea) for it attempts to do what most experiments have failed to do. It isolates the specific disabilities of industrial society in relation to the natural world by looking at the culture of some groups of people who have not been able to go back to the city when life became difficult.

It starts with some examples from the ethnographic literature of the Pacific islands to illustrate the kinds of relationship between environment, human beings and religion which have produced sustainable societies. The purpose is not to advocate a return to the past, which is impossible, nor to romanticise subsistence economies, but to see whether there are lessons to be learned from such societies which should and could be incorporated into the philosophy of a sustainable post-industrial society of the future.

The chapter then discusses the consequences of the contact which took place between indigenous peoples and the industrial civilisation which invaded their lands, mainly in the last 200 years. Even if Westerners as a whole were to become as disillusioned with modern life as the eco-activist minority and wish, as a whole, to return to the 'Garden of Eden', it is now unlikely that the descendants of those whose Edens were so recently despoiled would want to follow them. But, though the forms have changed, much of the substance of pre-industrial morality has survived our best efforts to change it, and might be built on as protection against the pitfalls which industrial society has failed to avoid.

Chapter Two

The real Garden of Eden

Throughout history there have been reformers who have pointed to some kind of 'golden age' in the past as a model for the present. Others, like the satirical Swift (1726) and the dead-pan Margaret Mead (1939), have looked sideways at contemporary or imaginary societies which had, it seemed, escaped some of civilisation's most famous discontents.

One writer who revived this tradition of social criticism as an advocate for the environment was Alan Moorehead, whose best-selling book, *The Fatal Impact* (1966), was as much a criticism of environmental damage as it was a lament for cultural devastation. The impact Moorehead wrote about was that of Western Europe and its value system on the human and animal communities of Oceania, including Australia and Antarctica. It is of particular interest to the modern environmental movement because it does not differentiate thematically between humans and other species. Men and women, like whales and seals, were assumed to have been living in a state of harmony with nature which was rudely shattered for ever by the first splash of the anchor of a European ship into the waters of an island lagoon.

The book had an important effect on middlebrow North America, Britain and Australia, and is said to have given President Kennedy second thoughts about the war in Vietnam, but it is an elegant expression of grief rather than a discussion of the principles of pre-contact environmental management, and no attempt is made to establish the basis of the innocence supposedly lost by the Tahitians or Aboriginal Australians. The difference between European and Pacific island society is seen as technological rather than ideological, echoing the assumption of Captain Cook (writing in 1777) that islanders were passive recipients of the agents of change:

I cannot avoid expressing it as my real opinion that it would have been far better for these poor people never to have known our superiority in the accommodations and arts that make life comfortable, than after once knowing it, to be left abandoned in their original incapacity of improvement (cited in Moorehead 1966, p. 40).

Moorehead assumed with Cook and his contemporaries that South Pacific societies were static. It was therefore consistent to deal with them in the same book and by the same thematic treatment as with other species. But in doing so he anticipated, perhaps unintentionally, the assumptions of those modern environmentalists who have stressed man's place in nature as that of a species among species (see, for example, Devall and Sessions 1985).

In fact, South Pacific societies were neither static nor were they passive recipients. They proved extremely selective in what ideas and technologies they borrowed from their visitors, and they often seemed perverse to Europeans in the uses to which they put them. Different kinds of land use, tenure and economic organisation at the time of contact are accordingly better understood as indicators of cultural preference or, in other words, the ideological relationship between a human society and its environment, than as indicators of positions on some hypothetical 'ladder of progress'. Those same cultural preferences remain an important ingredient of contemporary island society.

Oceania was remarkable for the cultural diversity which resulted from its insularity. It included the hunter-gatherers of Australia, the shifting cultivators of Melanesia and the agricultural societies of New Zealand and the larger islands of Polynesia, whose sizeable surplus had given rise to powerful kingdoms. The European perception of this environment was of nature untouched by the hand of man, a virgin territory ready for the taming hand of civilisation. The idea that the environment had not been altered by the indigenous inhabitants, who were 'part of nature' and thus seen as less 'human' than the invaders, was to become an important ingredient of nineteenth-century racism. It lingers on as an underlying paradox in the ideological inheritance of the modern ecology movement, which is anti-racist because racism has facilitated the expropriation of land and resources from indigenous peoples. However, the movement also tends to advocate the nurturing of indigenous cultures because of their assumed custodial superiority rather than because of their equal human value. It was not, primarily, the destruction of human cultures, but of the rainforests of the Amazon, Malaysia and Indonesia, which led to the recent green slogan 'save the tribal peoples', while Michael Mansell (1987), the Tasmanian Aboriginal leader, recently referred to conservationists who advocate the extension of national parks as 'trespassers on aboriginal land'. There are thus considerable obstacles in the way of a potentially powerful alliance between green politics and indigenous land rights movements. A less romantic appreciation of pre-industrial relationships with the environment on the part of environmentalists may help to overcome this difficulty. Thus, for tribal peoples, 'wilderness' is an invention of the European imagination. If it did exist in the sense of land totally lacking human inhabitants, visitors or users, its value, to tribal peoples, would be questionable. It was the assumption that there was such a thing as wilderness as an unclaimed 'commons' which led to the appropriation by invading colonists of land which tribal peoples never doubted that they 'owned', however variously that word may have been interpreted.

In fact, Oceania and Australia had been civilised in accordance with the cultural values and with the technology available to the inhabitants, and the environment had been greatly modified. New Zealanders, Tongans and Fijians had built huge docks and fortifications. Hawaiians had built artificial fish ponds and had terraced their mountain sides for intensive agriculture (Bellwood 1978, pp. 354–5). Solomon Islanders had built artificial islands and Fijians had dug waterways for navigation by ocean-going vessels. Aborigines had built dams on Australian rivers to irrigate pastures and hunting grounds and to conserve supplies of fish; they had introduced some new species to the continent and exterminated others. They used fire, regularly, throughout the Australian continent as an aid in hunting and for control of vegetation. Its constant use for thousands of years favoured the growth of grass and fire-resistant eucalypts in place of the formerly dominant casuarina forests, thus transforming the vegetation pattern of the continent (Blainey 1982).

'Natural man' has never abandoned the heritage of a distinctive humanity, or been loath to place his vital interests ahead of those of other species. Nor need we, but because of our disproportionate power, we are victims of a false dichotomy between the welfare of humanity and that of other species. Pre-industrial society was not able to make this mistake. Without the illusion of omnipotence, the interdependence of species was so obvious as to be enshrined in many forms of religious belief. Throughout Oceania and Australia, much of Africa and pre-Columbian America, the relationship between 'man' and his environment was one in which 'he' saw himself as a factor in the cosmos, not something beyond or outside it. Man sought to manipulate nature to his advantage, to utilise its forces, using the spiritual technology which defined his culture, but he did not think of mastering it. Knowledge of the natural order, whether in an Australian desert or on a Pacific atoll, enabled man to survive and to produce a surplus, from which could spring specialisation and leisure, music, theatre, oratory, warfare and politics without fundamental separation of man from nature. Indeed, it was 'man the sorcerer' rather than 'man the toolmaker' who, because of his sense of history and his spiritual prowess, could enlist the support of *natural* forces in his struggle for pre-eminence.

The nineteenth-century categorical distinctions between spiritual and natural, religious and secular, were introduced with difficulty to cultures in which the spiritual included, in some degree, everything which foreigners regarded as natural or secular. Missionaries were therefore often surprised by what seemed to be the literal-mindedness of their congregations who expected the same kind of concrete responses to prayers as to the cures and curses to which they were accustomed. Now, in the late twentieth century, our culture has once more narrowed the distinction between natural and spiritual by our belief that there is a scientific explanation for everything. This enlarges the natural sphere to embrace the spiritual. The result is that the views of a hardline scientific reductionist of the present day on the subject of man's obedience to the laws of nature are no less rigid than those of a pre-contact New Guinea highlander, though the highlander would be more likely than the scientist to act in accordance with his philosophy.

The dominant ideology of contemporary science holds that the destiny which shapes our ends is our biological nature, that man is no more than the sum of his parts, and that the laws which govern the interactions of the smallest particles of matter are mirrored by discoverable laws which govern human society and the way it relates to the environment. But outside their laboratories scientists, like other people, are the inheritors of a philosophy which gives man free will, which places him, moreover, at the centre of the universe, the sole spiritual creature for whom the world was made, and whose culture, that is, his relationship with the environment, is one which sees the biosphere as a resource for human use rather than an entity of which human society is a part.

The problem in working out an environmental philosophy for a sustainable modern society is not so much to adopt the technology of the New Guinea highlander, but to achieve the same degree of intellectual consistency.

To turn now to some examples. Polynesia and New Zealand got a good press in Europe in the early nineteenth century, because Polynesians fitted easily into the role of noble savage which had been prepared for them. The botanist, Sir Joseph Banks, lost his scientific detachment as soon as he got to Tahiti: 'the scene that we saw was the truest picture of an arcadia of which we were going to be kings that the imagination can form' (Beaglehole 1962, vol. 1, p. 252).

Australian Aborigines got a very bad press. Dampier thought them 'the most miserablest people in the world.' Cook and Banks found them gentle and timid, but thought them backward because they showed no signs of personal ownership. They were therefore placed within the eighteenth-century 'scale of being' at the bottom of the human heap, just above the animals.

When it came to contact with Melanesia, the similarities between European and Melanesian materialism were ignored and the descriptive emphasis of early travellers was on the primitive technology, the apparently unsophisticated and fragmented political organisation, the sorcery and witch-craft. Later writers have stressed the relative egalitarianism, the individual ownership of many kinds of property, the often 'puritanical' moral code. If Polynesian society was, in European eyes, an idealised re-creation of the pre-industrial society Europe had lost, Melanesian society was a caricature of the capitalist society which Europe, North America and Australia were busy creating. There is some sense, therefore, in examining the relationship between social organisation and environmental management in the context of such small-scale societies where causal relationships are rela-tively uncomplicated.

Status in Melanesia, as in most modern, affluent societies, is characteristically measured not by hereditary rank, but in wealth, and though wealth is measured in things like pigs rather than in corporation shares, it produces, as in America described by Charles Reich (1970, pp. 33-4), men who combine an ostensible interest in the welfare of the community with an economic calculation based on self-interest. They also have a heavy impact on the environment.

Melanesian 'big men' who bring new forest clearings into cultivation, and so place a wide network of kinsfolk under an obligation to them, have their parallels in the railroad speculators of the American West or the 'beer barons' of Australia, men who believed, with Adam Smith, that the pursuit of self-interest is often the best way to serve the community. But the 'big men' of the twentieth century operate in an almost totally secular context. The long process by which custom becomes sanctified by authority, institutionalised and codified as dogma, has been set aside, and with it the guarantee of the grain of truth which lay within the original paradox, because of the historical circumstances in which it was conceived. Societies with rituals still powerful enough to ensure the exercise of individual power in the interest of the community, while allowing scope for the exercise of entrepreneurial talent, should be of particular interest to those environmentalists who place a higher value on liberty than on equality, and therefore favour monetary imposts and incentives, rather than compulsion, as a means of making industry environ-mentally responsible.

One anthropologist, Roy Rappaport (1967) suggests that

the placing of a high value on sanctity is the corollary of having a low level of technology. At one end of the scale are societies governed by sacred convention with very little human authority in evidence. Next come societies which have very sacred authorities, but with little real power, like Tonga, with its sacred king, the Tu'i Tonga, and his administrator, the Tu'i Kanokopolu, who was perceived by European observers as a secular figure. At the other end of the scale are the industrialised totalitarian societies like Nazi Germany or Soviet Russia, where authorities have very little sanctity but a great deal of power. Rappaport argued that this continuum correlates roughly with technological development. It is this which gives men the power to manipulate the environment to their advantage, which enables them to dispense with sanctity. For societies lacking mechanisation, ritual becomes the technology of environmental management; taboo, or tapu, to use the original Polynesian word, becomes the means of conservation.

Rituals are actions designed to ensure the co-operation of spirits or gods in human activities, like planting crops, raising pigs, wooing women, building ships or waging war. They are the fulfilment of a contract by the living to which the spirits of the dead, or the gods, must respond. If, in spite of performance, crops do not grow, pigs die, canoes capsize, it is not because the spirits are recalcitrant but because the ritual was inexpertly or incorrectly performed.

Tapu means a state of sacred preservation. In Fiji, for example, the chief was the spiritual source of all well-being and fertility. The chief's head was therefore tapu, and other heads had to be always closer to the ground. This could be seen as a customary rationalisation of a precaution against assassination, but was accepted as a spiritual sanction. Women were tapu to their husbands after childbirth for varying periods in most island societies, which had the effect of securing adequate spacing of children. Temporary tapus were imposed on particular trees or on such things as spawning fish. Tapu provided a religious reason for refraining from anti-social or anti-environmental action and thus formed the basic ground rules of environmental management, community health and control of population. There are thus reasons for the assumption, which many have made, that environmental problems would be solved if the ideologies and technologies of pre-industrialised societies could be restored. The fact that this is impractical for most people in a modern society usually aborts further discussion, but it may be important for those seeking a philosophical basis for a sustainable society to consider, briefly, what 'living as if nature mattered' really means, for those who have no choice about it, so that they can be selective about applying the lessons of their experience to modern society.

The Tsembaga, who live in the highlands of the Madang district of New Guinea, provided Roy Rappaport with the material for his famous study of environmental management, *Pigs for the Ancestors* (1967). It has caused continual and heated debate ever since its publication and his material is not introduced here as an example to be emulated by environmental reformers but as an illustration of the logical conclusion of the 'back to nature' argument as a remedy for environmental problems. The technology is modified in this example by Western contact, and includes steel axes and bush knives alongside digging sticks and bows and arrows, but the Tsembaga have found their place in a view of the world and of the environment which sees humans as members rather than masters of the ecological system around them.

The country is mountainous, rising to 3000 metres and topped by virgin forest. This area is controlled by the red spirits, the spirits of ancestors killed in warfare. They forbid the felling of trees and they are the objects of rituals concerned with war. They impose the tapus which enforce periods of peace. The area they control is seen, not as a reserve which could be cultivated if the spirits would allow it, but as a resource in its own right, providing firewood, building materials and a livelihood for hunted game. The red spirits are envisaged as having the characteristics of heat, strength and hardness, and as concerned with the welfare of the top half of the human body. If someone gets sick in the head, pigs are sacrificed to the red spirits and eaten by the patient and his relatives.

The territory below about 1000 metres above sea level is often swampy and unhealthy, inhabited by malaria-carrying mosquitoes, but more importantly, by the *rawa mai*, the spirits of the low ground. *Mai*, the word for low in the local language, also means something out of which something else has grown, so the spirits of the low ground, the spirits of rot, are also the spirits of the whole cycle of fertility, growth and decay, which includes all living things. In contrast to the red spirits, their characteristics are coldness, wetness and softness. They are responsible for the lower part of the body and implicated in anything which goes wrong with it.

Most living is done in the overlapping area between about 1000 and 2000 metres above sea level. Scattered clearings are cultivated in this zone, which is thus one of secondary growth. Each clearing is cropped intensively for a period of twelve to fourteen months and then abandoned for fallow periods of forest regeneration of about twenty-five years.

Except for the years chosen for major festivals and periods of warfare, when people move into temporary settlements, people live near their gardens. It is an agricultural system which achieves great economy in the use of energy, though it is less efficient than more settled kinds of cultivation in terms of production per acre. There are no transport costs of production, however, and it provides not only subsistence, but a surplus for feasts, entertainment, decoration trade and warfare on the basis of an average working week of about twenty hours, provided the correct rituals are performed.

The gardening practices themselves have developed in a hot and humid climate with conservation of energy as a major objective. Weeding is done selectively, and particular species of young trees are allowed to grow up between the root crops as a means of placating the spirits of fertility. This is a spiritual strategy which ensures that the same land is not cropped for too long. Gardens are eventually abandoned, not ostensibly because the fertility of the soil is exhausted, but because the developing trees reach a size which makes it difficult to harvest the remaining root crops. Pigs are then placed in the abandoned garden for a few weeks. They chew up the small plants and root out the unharvested vegetables, which gives the young trees a good start.

Domestic pigs are the main source of protein, but they are kept for special ritual occasions. The marginally adequate source of protein from day to day is supplied by hunting smaller animals and occasional wild pigs in the forest. Protein consumption is restricted and directed to the people who most need it. If men, for example, eat wild pigs, it is believed they will get lice in their hair. If pre-pubescent girls eat rats they will smell bad and grow up to be unattractive.

Young children and women who have children to look after have no tapus on what they can eat. Adolescents need protein, but do not get as much as they want. This seems to delay sexual maturity, and combined with a prohibition of sexual intercourse for males during hunting seasons and for long periods leading up to warfare, it seems to keep rates of human reproduction constant.

The main function of domestic pigs is to serve as 'energy banks'. They are useful in small numbers as aids to agriculture. They live around the houses and clean up living areas, thus converting low-quality vegetable food into high-quality protein which is used in times of stress. The major use of pigs is in the rituals associated with war; they are sacrificed to the red spirits to gain their military cooperation in the next round of warfare against traditional enemies.

But they are also sacrificed, and eaten from day to day in cases of accidental injury or illness, to either the low spirits or the red spirits, depending on which part of the body is in trouble. The liver goes to the victim, the remainder to close relatives, the men getting the fatter parts, the women the lean. Rappaport argues that the effect is to promote healing in times of crisis and protect those most exposed to infection. The pig population becomes a barometer of the suitability of a particular place for human habitation; thus a 'good' place is defined by the Tsembaga as a place where people do not get ill very often. The pig population therefore expands because there are few demands placed on it. A 'bad' place is a place where there is a lot of sickness and so the pig population only increases slowly.

It is a common saying in Melanesia that the more wives a man has, the more pigs he can keep, because the women do the gardening which produces food for the pigs. A 'big man' therefore starts his career of upward mobility by multiple marriages which in turn enable him to raise pigs enough to entertain on a grand scale at the appropriate time and so keep the living as well as the spirits of the dead in a state of indebtedness to him for service to the community.

In this part of New Guinea, the relationship between

women and pigs is the controlling factor in the ritual cycle which in turn controls the relations between human beings and the environment. The crucial factor is the number of pigs which one woman can provide for. At first they are treated very much as domestic pets, but a change then takes place from a relationship of support, in which pigs serve a useful function as scavengers around the house, and as emergency supplies of protein, to a relationship of parasitism, when the owner has to begin to plant extra food and work longer hours to keep the animals. This change takes place when the ratio reaches about four pigs to one woman.

The ritual cycle begins after a period of hostility is over. A truce ritual takes the form of planting a particular kind of small tree called a *rumbim*, which is believed to demonstrate the successful defence of the tribal territory. At this point, the people stand in debt both to their allies and to the spirits for their help in the recent fight. The truce therefore lasts until the pig population increases to the point when there are enough to present to the spirits and ancestors to secure their support for the next round of hostilities. In the meantime, the truce is symbolised by the *rumbim* trees. The territory cannot be defended while they stand in the ground and the ritual of rooting them out cannot take place until the pig population has recovered.

Left to themselves the pigs would multiply to the point at which environmental degradation would begin, but the insurance against this is the fact that the physical capacity of the women to care for the pigs is below the capacity of the territory to provide the animals with sustenance. So, after a few years, the women begin to complain to their husbands about the fact that the pigs are starting to break down the fences around the gardens, causing fights, increasing social friction and making more work for everyone. In a 'good' place, where disease is rare and pigs multiply, this may take only five years. In a 'bad' place it may take fifteen.

The men eventually respond to domestic pressure and discuss the staging of a year-long festival known as the *kaiko*. In this relatively egalitarian society with no hereditary rulers, discussion has to be continued until eventually a consensus is reached. This happens when domestic friction for a majority of husbands becomes intolerable.

The *kaiko* then begins. There are five stages. The first is the planting of boundary stakes. The second is the removal of settlements from garden areas to a central dance ground. The third is the ceremonial uprooting of the *rumbim* trees by the men who planted them, or their sons if they are dead. The fourth is a dance, to which potential allies are invited, and women and trade are exchanged. Finally, there is a night-long dance where tapus are removed, signifying that war can now commence.

Each stage is noted and balanced by customary enemies within the region and is also marked by the large sacrifice of pigs until the point is reached at which the spirits are in debt to the living and their support in the next round of warfare is guaranteed. The effect is that the population has been on a high-protein diet for the previous twelve months and dancing and other vigorous activity has raised the general level of fitness.

The total effect of the ritual cycle of between five and fifteen years is thus to limit the seriousness of conflict between humanity and the environment and between human communities. It adjusts ratios between land and population, facilitates trade, and distributes surplus pig populations throughout the region through the giving of feasts. It limits fighting to frequencies which do not endanger the existence of the regional population even if local communities suffer large casualties.

It prevents the accumulation of surplus, however, to the point at which it could give rise to more valuable cultural achievements, by ensuring the repeated renewal of inter-communal hostilities.

Such societies may command ecological admiration, but their ideological rigidity and their violence makes them less attractive as models for emulation than the fanciful and peaceful tribes of imagination. This is not just because most modern eco-activists lack the hunting and gardening skills of pre-contact Pacific islanders and would find the living conditions uncomfortable, but for some theoretical reasons as well. The most important is the difficulty of providing an equivalent of the particular and often locally specific beliefs and epistemological systems which provided a basis for the sanctions and constraints which held such societies together and kept them functioning. The Australian Aboriginal Dreamtime, for example, both explained the particular

landscape to the limits of experience and in doing so defined
both social and ecological obligations. Without an equivalent to the Dreamtime, incorporating responsibility *to*
particular ancestors and *for* particular descendants, as well
as contemporary totem species and human kin, too many
people will break the rules.

Such equivalents do exist. Edmund Burke, in England, was
discovering for himself the importance of intergenerational
responsibility as an antidote to revolution at just about the
time that Cook came upon societies which lived by such
principles in the Pacific. But in modern times, it is only
those communities which have been bound together by a
common experience of persecution, religious fervour or a
combination of both, that have maintained at least partial
ideological consistency. And their purpose has been to
provide an alternative to mainstream society, not to change
it.

A powerful ingredient of success in those societies which
have established the most successful relationships with the
environment is an emphasis on kinship, involving unbreak-
able bonds of obligation, specifically defined, to a large
network of relatives, both living and dead. Kinship is thus a
means of defining social function and ensuring intergenerational and communal co-operation. Many eco-activists are
strongly influenced by anarchism and libertarianism, and see
that concern for property and patriarchy which has characterised Western European society as a major cause of the
environmental crisis. Others are in flight from what are seen
as the too exacting restraints of even the nuclear family,
which is all that remains of what has been, in all societies,
an important means of nurturing a sense of intergenerational
responsibility among its members. Those who seek a consistent environmental philosophy might therefore need to re-
evaluate the importance of the family, with its obligations
to particular children and particular parents, as likely to
provide a more compelling rationale for good environmental
management than individualism tempered by an abstract
notion of intergenerational responsibility. Environmental
responsibility in tribal society is not usually an abstract idea
but a function of extended family responsibility in which
continuous blood ties are periodically reinforced by rituals
and ceremonies. It is difficult for bonds of such power as this

to develop in the context of a voluntary democratic community held together by intellectual ties alone, especially when the motivation for communal life is not so much the welfare of society, but, as it often is, the rediscovery of self (Munro-Clark 1986, p. 33).

But if there is little that can be borrowed for immediate practical purposes from the tribal societies of the past, there may be something to be learnt from the selectivity they exercised when confronted by the invasion of the West, and by their current response to the more recent assaults, symbolised for many by the nuclear explosions on Bikini atoll in Micronesia, Moruroa in Polynesia and at Maralinga in South Australia. The significance of the enquiries into these events arises from the context of the reassertion of Aboriginal and Polynesian cultures in which the debates about nuclear testing take place. It is a development which Moorehead (1966) did not anticipate.

Since Moorehead wrote, research by historians such as Dorothy Shineberg (1967) and Kerry Howe (1974) has accounted for the cultural resilience of island populations by showing that the introduction of muskets by Europeans was by no means as devastating in its effect as might be supposed (Howe 1974). The damp world of open boats, salt water and tropical forest, was not an easy one in which to keep gunpowder dry. Islanders soon had access to all the arms they wanted in any case, and they tended to be absorbed into traditional trading systems without greatly altering the goals for which wars were fought, or trade took place, or the relationship between the environment and society. The Aborigines of Australia, however, did acquire firearms in any quantity until the twentieth century because they were not in a position to trade with the invaders. By the mid-nineteenth century, Europeans had a great military advantage because of the introduction of the breech-loading rifle which, unlike the musket, was accurate over long distances. It made it impossible for dispossession to be resisted successfully and so it was in the period from 1840 to 1870 that the management of the Australian environment fell almost exclusively into the hands of the invaders.

While the effects of the introduction of firearms varied, the introduction of European disease was universally devastating. Individual cases of illness posed no problem to indigenous societies. Sickness was of spiritual origin, a necessary prelude to death, and hence part of the cycle of fertility and renewal. It could, of course, be manipulated and its effects modified, like other aspects of the natural environment, through the use of correct ritual.

While deaths occurred one at a time in small communities, there was nothing to disturb this understanding. They could be individually attributed to the malevolence of a particular god, ancestor or sorcerer. Epidemics, on the other hand, posed an immediate challenge to the whole epistemological system, since they resulted in the deaths not just of individuals, but whole age groups and sometimes entire communities. Venereal disease troubled the conscience of Captain Cook and enhanced the image of lost innocence, but it was the epidemics of dysentery, measles, smallpox and influenza which were mainly responsible for the cultural devastation of Oceania. Smallpox, transmitted to Aborigines from contact with explorers, sealers and whalers, is now believed to have greatly reduced the population of Australia well before formal settlement. Repeated epidemics of dysentery hit Fiji from the end of the eighteenth century, to be followed by measles in the late nineteenth century and influenza in the early twentieth century. It was a typical sequence. The effect in the case of most indigenous societies was literally to de-moralise them, that is, to undermine traditional beliefs and so to deprive them of a rational basis of morality. It is argued from some recent research in Fiji that when the gods evidently became unpredictable, cruel and capricious, chiefs and commoners followed their example. It meant that the first Europeans to encounter them were dealing with traumatised society (Heasley 1982). The lesson here is that the creation of a moral vacuum, when societies abandon the rules which normally apply to human relationships, leads to the additional loss of a rational basis for a sustainable relationship with the environment. The death of Lake Baikal and the nuclear pollution of the Pacific are the products of amoral societies. Similarly, in the Pacific islands, species like sandalwood became locally extinct within as little as a decade in answer to the demand for Western trade goods. It was not that European goods were overwhelmingly superior; often they were rejected as unsuitable. The problem was that chiefly authority became ineffectual. Originally authority was derived from the chief's role as environmental manager, the source of fertility and well-being. Cakobau, high chief of Fiji, is believed to have described the traditional role of the chief to Sir Arthur Gordon in words which not only express a traditional concept, but also anticipate the ecological view of the responsibility of government in the 1990s. 'The land and the people', he told Gordon, 'are one. We rule both, but we own neither'.

The understanding of health and sickness, population and resources, political authority and environmental management, in any society, is likely to be logically consistent. American and British concern with access to strategic resources locks them into the support of an oppressive racist regime in South Africa. Societies which spend more on curing the diseases of affluence than on avoiding the diseases of poverty are likely to favour things like uranium mining and denying self-determination to indigenous peoples. If one set of views is undermined by experience, popular attitudes in the other main areas are likely to change too. So it was with the conveniently small-scale societies of the Pacific islands, convenient, that is, for analysts of social change.

The existence of a moral vacuum, created by the sustained social disasters of epidemic disease and the undermining of chiefly authority, provided an opening in the early nineteenth century for those missionaries who preached that disease was God's punishment for sin and that conversion would entitle the victim to Western medicine, the technological fix of the 1830s. The doctrine often backfired. The great advances of antiseptics and anaesthetics, let alone antibiotics, were still a long way off. Missionaries usually knew much less about the human body than their cannibal patients, and medical science itself was at the stage when the grim alternatives of recovery or death were often still the preconditions of diagnosis. Western medicine did, in some cases, induce conversion, but it remained an optional and occasional alternative to traditional medicine, and the identification of missionaries as medical practitioners meant that they reinforced the existing sociological identity between doctor and priest, and of priests as the proponents of a particular technology. Missionaries fell into more traps as they sought to control the environment, praying for safety in storms, thanking their God for fruitful crops, growing church funds and victory in war (Cowen 1985).

By the last quarter of the nineteenth century, island leaders of the Pacific, now dominated politically by Europe, were discovering that, having lost its evangelical steam, Christianity had become, like its predecessors, an indigenous institution. It was able in many cases to pick up the pieces of a fragmented culture and become the ideological core of a reaffirmed identity. Tonga was the first island society to establish a specifically national church, in 1885, closely identified with the Tongan monarchy. Elsewhere, as in Europe three centuries before, church leaders took the lesson to heart and indigenised local church leadership, which had the same effect.

Michael Somare once pointed out that one of the differences between trade unionism in New Guinea and in other countries is that in New Guinea, trade union meetings are opened with prayer; there is thus once more a rejection of the division between religious and secular spheres. Religion remains, as it always was, a technique for social and environmental management. Elsewhere in the region, societies have found different paths to the reassertion of cultural identity, moving in some cases from traditional beliefs, perhaps to Christianity, sometimes back again in response to crisis, through a cargo cult, or, like many present-day Aborigines, a pentecostal movement. These movements can perhaps be understood as the development of a series of options in response to crisis until an option is found which succeeds, as all successful religions must, in fulfilling secular needs. These needs are, for the survivors of the post-colonial world, a rational base which reaffirms cultural identity in the context of underdevelopment.

For this reason, the most enduring religious movements have been those which have facilitated a re-establishment of the bond between people and land. The Ratana Church of New Zealand, the Wesleyan Church and the Taukei movement in Fiji, the Methodist Church and its leader, Rotan Tito in Banaba, the Pentecostalists of Arnhem land, the Kanaka Church of New Caledonia, are obvious examples. The common ingredient, as with the Israelites and of many migrant minorities displaced by the imperialisms of other times, is the concept of a promised land or a land to be retained or returned, the precondition of the implementation of indigenous environmental philosophies.

Europeans, observing this cycle of reaction and response have usually misconstrued the evidence, and may miss the implications for the dilemmas of the present and of the relationship between health, human relationships and the environment.

It seemed at first, in the early decades of contact, as though the epidemics were the beginning of an apparently irreversible process. Edward Markham, a visitor to New Zealand in 1833, spoke for his contemporaries when he ventured the view that

Throughout the world, the same causes may be seen at work . . . Rum, blankets etc., have been the great destroyers. But, the Almighty must have intended it to be so, or it would never have been allowed to happen — out of evil comes good (cited in McCormick 1963, p. 83).

Perceived first as a vindication of religious belief, it was not long before Darwin's work was assumed to provide such theories with secular scientific support. The decline of native populations was therefore accepted, if not with gratitude, at least with complacency, but by the end of the nineteenth century it was, for a mixture of reasons, becoming a matter of regret, and colonial administrations were trying to do something about it.

A typical exercise was the Fijian enquiry into the causes of native depopulation conducted in 1895, which collected a vast amount of evidence and opinion, mainly from Europeans. The commissioners stressed the devastating effect of the early epidemics, but the consensus of evidence was that the remedy was to assist native peoples to 'catch up' with Western society as much and as rapidly as possible. This meant becoming essentially extractive in their relationships with the environment. Living standards should be 'improved' through the development of cash crops, the individualisation of land tenure (for then it could be made subject to economic laws, perhaps sold to Europeans) and the introduction of incentives, anticipating, in many ways, the views of the development economists of the 1960s. Depopulation was caused, it was believed, by the people 'subsisting upon the stumps of their old customs', by which they meant such vestiges of pre-Christian culture as abortion and infanticide, but also the communal system itself, and the authority of the chiefs, which seemed to stand in the way of the development of Fiji from a semi-feudal society to that higher stage on the ladder of progress, an independent yeomanry. Fijian mothers were accused of neglect, and disease was attributed to lack of Western medical attention, predilection to 'quack' remedies, and traditional styles of housing (Fiji Government 1896).

The colonial administration began to train native medical practitioners, islanders with an introductory education in Western medicine, taught to administer a simple range of drugs and dressings. The cash side of the economy was encouraged, as was labour on European plantations, so that Fijians could purchase sawn timber and corrugated iron for the construction of supposedly hygienic houses, and eat imported tinned food and biscuits. Fees could be paid for the more promising young schoolchildren to obtain a Western education and so lead the movement away from the communal system and towards the work ethic. It was not a change which could be brought about overnight, but the need was urgent. In the words of one District Commissioner, T.R. St Johnstone (1913):

I doubt whether one, or even two generations will see a nation of careless children changed to a nation of workers . . . And they will have to be quick, for the childlike races cannot survive the stress of modern competition . . . Only the strict rules and discipline of schools on the European system, acting promptly on the rising generation, can save them, as a race, from complete moral decay.

In the case of Fiji, the population as a whole continued to decline until the mid-1920s, in spite of the government's best efforts. However, in Lau, the eastern islands of Fiji and St Johnstone's district, an increase first became clearly evident in 1907. For the Commissioner, the reason could not be the greater extent of Westernisation, for there was no evidence of that in this remote and isolated group of islands; he fell back on popular racist mythology: 'the people in western Fiji are much lower on the scale of civilisation, have a different type of dwelling house, different features, and are altogether of a more Fijian type than the Lauans' (St

Johnstone 1910). The Polynesians of eastern Fiji were, he believed, infinitely superior to the Melanesians of the West. This explained, he believed, why Lauans had responded well to educational opportunity, though there was still a long way to go, especially in the area of what he called domestic hygiene: 'The women of Lau are still steeped and hidebound in the old native customs and superstitions, many of which are hygienically disgusting and the cause of the decay of the race' (St Johnstone 1914).

In fact, the race was not decaying at all, and its numerical recovery had nothing to do with Westernisation. It was soon noticed that it was in the most remote islands of all, whose isolation had made the retention of the traditional relationship with the environment essential, and whose subsistence economy still flourished, that births most greatly exceeded deaths. The discovery made it difficult to sustain the belief that the population would only recover if Fijians abandoned their traditional way of life. Better nutrition provided by subsistence agriculture and fishing than by the penetration of the market economy in other areas may also have been part of the explanation.

By 1918 the trend had become too clear to ignore and it was difficult to account for in terms of contemporary development theory. William Sutherland (1918), who had now succeeded St Johnstone, reported:

The usual large increase in the native population of Lau has not failed this year. This yearly source of gratification to us all is due, in the writer's opinion, not to the branch hospitals, which are so amateurish as to be of doubtful utility, nor to any special zeal or cleverness of the native practitioners or other officials, nor is it due to the lack of any of the ordinary diseases which are endemic in all parts of Fiji ... Actually, the yearly increase is due to the work of nature, not of man.

He went on to attribute it to the quarantine effects of isolation which had protected the remoter islands from both the immediate and long-term demographic effects of the epidemics. It was a conclusion which was to be borne out by later more comprehensive research covering the whole Pacific islands region (McArthur 1968).

The Aboriginal populations of Australia, with those of the Pacific and parts of Africa, now have the fastest growth rates in the world. It is a development which is explained by demographic factors rather than, as is often supposed, the application of Western medical technology, which deals increasingly with the ills of urbanised, affluent society and requires ever larger amounts of capital to succeed in its goals. The effects of population recovery, however, have included instances of vigorous cultural recovery.

Cultural recovery demands at least two things: people and land. Culture is the relationship between them. Where indigenous peoples have been separated from their land as in the settler colonies of the temperate zones — South Africa, Australia, North America — a large degree of cultural homogenisation has taken place. Though even in some of these areas not all has been lost. The retention and recovery of Maori lands, the Aboriginal land-rights movement and, for all their racist basis, the Bantustans of South Africa, have provided the essential ingredient for a re-emergent identity. Elsewhere in the tropics, white settlement has never been more than temporary, and land use has been the basis of the survival or revival of indigenous culture. In China, though not to the same extent in South Asia, land reform has facilitated adequate subsistence and, more recently, the production of an agricultural surplus, as opposed to an agricultural substitute of cash crops for food. In Africa this did not happen, and the pattern was that of colonial mono-culture characterised by what has been called the 'dialogue of the dumb', in which World Bank and FAO experts spoke only to a bureaucratic elite in air-conditioned buildings about problems which neither had ever experienced at first hand.

Oceania provides a range of examples which illuminate the problem and may suggest the kind of philosophical reorganisation which may be needed on a world scale if environmental management is to succeed in the goal of sustainability.

The pattern of resilience and recovery from the 'fatal impact' has mirrored the extent to which indigenous peoples have been able to hang on to their land in the nineteenth century. That is to say that peoples who have succeeded in retaining most of their land, whatever else they may have lost, have experienced a minimum of cultural disruption. Those who have lost most, like the Australian Aborigines,

have been the most culturally devastated. These consequences demonstrate the relationship between cultural
integrity and good environmental management. Maoris
retained some land, after a long war, and have bought back
more of it since 1916, when a Native Lands Trust was set
up for the purpose. The Maori King movement, and later the
Ratana Church, became the focus of a new nationalism
which modern Maori leaders have been able to build on.
Maori lands, which fall well below the national average in
terms of butter fat production per acre and other indices of
capital-intensive production, also have fewer problems of
environmental management. Their streams are less polluted
by phosphates. The regrowth of bush, which is sometimes
regarded by neighbouring Pakeha (white) farmers as evidence
of neglect, leads to less soil erosion and provides a
sustainable resource for the future.

Apart from the areas devoted to growing sugar in Fiji, and
smaller plantation areas in Samoa, Fijians and Samoans
retained most of their land, together with their subsistence
cultivation methods. Cash cropping provided supplementary
income when prices were favourable, but the copra industry
has not yet involved the introduction of heavy machinery, or
the use of large quantities of pesticides or fertilisers. The
imported luxuries of one generation have become the
psychological necessities of the next, but it is not yet
necessary to concentrate on a cash-crop monoculture, to the
exclusion of subsistence, in order to meet repayments on
imported goods. Hurricanes, which are an unavoidable
feature of the climate, persistently devastate cash crops, but
do much less damage to traditional food crops. Useful as
they are in attracting international aid, hurricanes are the
cause of much distress, but hardly ever of starvation. They
occur in Fiji, on average, every two years. They are part of
the regular climatic pattern and they are only exceptional
events in the sense that their occurrence cannot be predicted
within a given financial year. Any costing of the copra
industry on a sustainable basis must treat hurricanes as
normal and recurrent rather than as exceptions which can
therefore be discounted. Hurricane relief funds are in fact a
regular subsidy to the copra industry.

In these countries the corollary of the retention of land has
been the survival of a 'land ethic', together with a system of

subsistence and a social philosophy which makes an interesting comparison with the ideas of those environmentalists who claim affinity with 'the ancient earth wisdom of native peoples such as Chief Seattle' (Seed 1988, p. 13), because of both the common ground and the differences.

Fijian scholar Asesela Ravuvu points out that the Fijian word for 'tribe' is *vanua*, which also means 'land':

The people are the *lewe ni vanua* (flesh or members of the land) . . . They are the social identities of the land, and also the means by which the land resources are protected and exploited for the sake of the *vanua*, the people and their customs (Ravuvu 1983, p. 76).

Identity with the land is highly localised, and extends to the social system and the patterns of obligation and behaviour towards the living and the dead which tend towards the maintenance of a stable relationship with the environment.

The term, *vanua*, has physical, social and cultural dimensions which are interrelated. It does not mean only the land area one is identified with, and the vegetation, animal life and other objects upon it, but it also includes the cultural and social system (Ravuvu 1983, p. 70).

Apart, perhaps from a certain absence of individual freedom and scope for spontaneity, there is some affinity between these principles and those of deep ecologists such as John Seed, but there are also important points of difference. It would be difficult to argue that Fijians or Samoans did not place the interests of humanity ahead of those of other species, indeed the archaeological record shows that several non-human species have been exterminated in the relatively brief period in which this part of the world has been settled by the human species. Also, the corollary of the fact that not only every physical feature, every piece of land and every reef and lagoon has owners, however light their touch, means that there is no concept of 'wilderness' as understood and valued by modern conservationists. As Ravuvu (1983, p. 76) points out

For a *vanua* to be recognised, it must have people living on it and supporting and defending its rights and interests. A land without people is likened to a person without a soul. The people are the souls of the physical environment. Like the interdependence of the body and the soul, the people control and decide what happens to the land. However, the people cannot live without the physical embodiment in terms in their land.

Though contact with other cultures, and the economic and educational opportunities offered by industrial society, have widened the choice of people who have retained such values as these, 'subsistence affluence', as some writers have called it, remains a viable option where European settlement, cash-crop monocultures and commercial media have not destroyed it (Fisk 1974).

In Australia, unlike New Zealand, Fiji or Samoa, there is no legal basis for settler occupation of Aboriginal land, other than the failure of Aborigines to prevent it. Since dispossession, monocultures, and production for urban and overseas markets, have become all but universal.

This kind of land use, introduced in response to the markets of industrial Europe, depends for its success on climatic regularity, which is even more elusive in Australia than in the tropical Pacific islands. Averages, whether of annual rainfall or monthly temperatures, are not typical. The only certainties in the Australian climatic experience are extremes of flood, drought, fire and storm. Farming practices, though utilising larger and larger holdings, using land more intensively, clearing more of it, and investing ever larger capital sums on more and more sophisticated equipment, have not resulted in sustained higher yields. The farming community has often been unable to survive the remorseless series of calamities which befall it, year by year, without enormous government assistance. Even with this assistance and the physical removal of most of the island of Banaba and much of Nauru to fertilise the continent, production per acre is no higher, on the average, than it was in the 1870s. The soil often blows away, under monoculture management, more rapidly than it is formed, and it will be increasingly difficult to hold at present levels.

Somewhat stricken with guilt at the implications, for Aboriginals, of the impending celebration of Australia's bicentenary of colonisation, a group of intellectuals formed an organisation in 1976 to make a treaty between the Aboriginal people and the Australian government, which would recognise the prior ownership of the continent by the

indigenous people and seek concessions in the shape of land
rights to present-day Aborigines, in order retrospectively to
legitimise European occupation for the past two centuries.
Dr H.C. Coombs, one of Australia's leading authorities on
both environmental and Aboriginal affairs, was one of those
originally involved. He tells a story with characteristic
humility about a discussion he once had about the proposal
with a native of the Northern Territory:

'How long have you white fellas been here then?'
'About 200 years.'
'A lot of peoples have come here for as long as that. They've come,
taken what they wanted, and gone home. We Aboriginal people
have been here for 40,000 years, and when you white fellas have
been here a bit longer maybe, and you've taken what you can get
out of it, you'll be gone too, and we'll still be here. Why do we
need a treaty?

A brief 200-year experiment of industrial land use in
Oceania is thus a paradigm of an important theme in the
relationship between the rich and poor peoples of the world.
Monocultural land use in wealthy countries is only
sustainable as a result of the use of non-renewable resources,
both locally and in other parts of the world. The products of
the process — wheat is currently a good example — are
increasingly difficult to sell to poor countries which, though
hungry, have been impoverished by the same kind of thinking resulting in the same kind of policies. They are having
similar difficulties in selling their bulk commodities, such
as sugar, copra, timber and many minerals, for adequate
prices. Technological innovation has evidently been
misdirected for short-term economic reasons into what has
proved to be a dead end.

But neither pre-industrial agricultural technology, nor the
land-management techniques of hunter-gatherers, would
now be adequate to the task of satisfying the material needs
of the enormously increased and increasing populations of
the world either, so we cannot simply go back to them. A
sustainable use of the planet will require an investment in
diverse alternative forms of production, both for local
subsistence and export, comparable to the investment which
has been made in the overproduction of mass commodities.
William Clarke, an anthropologist, geographer and

research colleague of Rappaport, points out that few persons now caught up in the 'neotechnic' world would want to become palaeotechnic shifting cultivators. He recognises that when palaeotechnic man learns of the neotechnic world he generally wants to join, a warning against romanticisation of the palaeotechnic. But he argues that there is a structural difference between permanent and impermanent systems of production which he seeks to identify. 'A new world must begin with a new mind; if the image is strong enough, our successors will be able to work out the details as they go.' He anticipates some features, however, of a 'para-primitive society'. It would have a lower material standard of living, as measured by consumerist standards, in exchange for unpolluted food, air and water. Population, as in most tribal societies, would be controlled. Society would be socially and politically decentralised and its educational system would emphasise the place of humanity as part of the ecosystem, regardless of its level of technological sophistication. Agriculture would be diversified and would cater first for local nutritional needs, then for distant markets, thus preserving a variety of systems and cultures. 'We must disintensify', he says, 'leave a few untended corners' (Clarke 1977, pp. 377-81).

Fortunately for such visions of the future, the impact of the West was not fatal after all, and indigenous cultures will have an important contribution to make to the kind of intellectually consistent ecological wisdom which will be needed to underpin the world-wide task of environmental repair, restoration and maintenance. The genius of those peoples who have retained not only some of their land but also their ethos has lain in their ability to select, from the experience of Western invasion and colonisation, those aspects of our value system which reinforced their own, such as stewardship, reciprocity, family loyalty, community service and charity, but to reject such goals as the accumulation of individual wealth as an end in itself and ideas such as the separation of humanity from the natural world, which industrial society had developed.

The incorporation of this experience into a global environmental ethic is unlikely to happen in a hurry because unless there is an overwhelming catastrophe, such as a nuclear war, majorities in the northern hemisphere are unlikely to be

convinced that they have much to learn from the South
Seas. It may happen gradually, and, if it does, it will be
important to learn the lessons of particular responsibility
rather than to assume the existence of a homogeneous
'Fourth World' with a sentimental rather than a pragmatic
relationship with nature. In the meantime, an easier starting
point for the establishment of a sustainable society may be
a reconsideration of the spiritual and philosophical relations
with nature which are part of the cultural inheritance of
Europe.

Suggestions for further reading

The process of interaction between industrial and
subsistence economies in the Pacific region is richly
documented. A good starting point is the superb 3 volumes
of the *Journals of Captain Cook*, and several of his officers,
edited by J.C. Beaglehole (1955–67). The same editor's
Endeavour Journal of Sir Joseph Banks (1962) deals with the
first voyage to Tahiti, which Beaglehole describes as 'a
cornerstone of the Romantic Movement'. Missionary
sources include *History of the Tahitian Mission 1799–1830*,
by John Davies, one of the participants, edited and introduced by Colin Newbury (1961). I have also learnt much
from the papers of R.B. Lyth and David Cargill (missionaries
to Fiji) in the Mitchell Library, Sydney.

Analysis of the contact process which is relevant to this
discussion is to be found especially in H.M. Wright, *New
Zealand 1769–1840 Early Years of Western Contact* (1959)
and H.E. Maude in *Of Islands and Men* (1968). An
understanding of the functioning relationship between
particular 'tribal' societies and their environments can be
obtained from many sources. Those I have found most
useful are Stephen Phelps, 'A Study of Valuables', a Ph.D.
thesis from Cambridge University, Asesela Ravuvu, *The
Fijian Ethos* (1983) and *The Fijian Way of Life* (1984).
Marshall Sahlins' collection, *Islands of History* (1985)
contains highly illuminating insights relating to the contact
process. Tim Bayliss-Smith and R.G.A. Feacham's collection
of papers *Subsistence and Survival: Rural Ecology in the
Pacific* (1977) and the recent *Islands, Islanders and the*

World edited by Bayliss-Smith and his colleagues (1988) contain much useful material on contemporary island societies and their eco-systems.

Roy Rappaport's *Pigs for the Ancestors* (1967) was followed by a series of studies by others on the Maring peoples of New Guinea, some of them critical of Rappaport, who replied to his critics generally in *Ecology, Meaning and Religion* (1979). Brown and Buchbinder's collection *Man and Women in the New Guinea Highlands* (1976), William Clarke's *Place and People: An Ecology of a New Guinea Community* (1971) and his essay 'At the tail of the Snake' (1980) are among the most interesting works on this well-documented area.

Modern communes are dealt with in the American context by B. Zablocki, *The Joyful Community* (1971) and *Alienation and Charisma* (1980). Margaret Munro-Clark's *Communes in Rural Australia: The Movement since 1970* (1986) is sympathetic, but not uncritical, while one of the most thorough discussions of the joys and problems of communal living and alternative family structures in both rural and urban contexts is *Living Together* (1980), communally edited by Dorothy Davis and others.

Dr H.C. Coombs has written recently about the possibilities for an Aboriginal treaty and parallels between the contact experiences of Maoris and Aborigines in his 1988 Boyer Lecture, published by the Australian Broadcasting Corporation.

The idea that industrial society can in some way recover the lost innocence of the past or of contemporary non-industrial societies is seductive, but for society as a whole it is a dead end, so we must return to the crossroads again. This is not to say that nothing can be salvaged from the exercise. The indigenous inhabitants of the southern continents and islands proved more culturally resilient than was expected, or perceived, and they may succeed in carrying over the most useful aspects of their pre-contact ideology into the post-industrial future which they will share with the rest of the world.

Industrial societies are in some ways less fortunate. Individual groups of people can submit selectively and experimentally to the ethos of an organic relationship with the environment, but most are unable to accept the idea of death as the precondition of fertility and renewal, and are apt to take whatever steps modern technology can offer to avoid it. For this kind of reason — the appeal of modern dentistry as against incurable toothache, for example — the nation-states which constitute industrial society will be unlikely to return voluntarily to the simple life. If such a return occurs, and it may, it will be the consequence of catastrophe rather than democratic decision-making.

Having then eaten irrevocably of the fruit of the tree of knowledge, Eden is no longer a possibility. But what about going to our own ideological origins to see if something can be salvaged in that direction! It has clearly led us into a lot of trouble with the environment, but it consists of several strands. Some of them may provide starting points for the establishment of a sustainable society capable of the levels of productivity now needed to feed the world's population and meet its compassionate needs.

The Western tradition has certainly provided humanity with a justification for its separation from the natural world, but the spiritual and philosophical relations of Western pre-industrial society also contain the idea of the good husbandman with responsibilities for the transmission of an undiminished inheritance to future generations.

The next chapter starts with a discussion of biblical and Christian religious ideas, and the elements of Greek thought which justified an anthropocentric view of the universe, modified by the concept of 'good stewardship'. These ideas were the ideological framework of the transition in Europe from various subsistence societies to one which was characterised by a settled agriculture, limited urban development, and considerable trade, but was still seen by contemporaries as being in a 'steady state'.

Economic growth in Europe from the twelfth century onwards, which accelerated rapidly from the fifteenth century created the *need to seek theological justification for the achievement of an apparent mastery over nature. Some secular thinkers moved more rapidly in the same direction, being less constrained by authority, but humanitarianism and liberalism contained important counter-currents. Concern for slaves and factory workers led to concern for animals and called into question the right of owners and masters to do as they liked with their own. Malthusian pessimism was the ancestor of the modern populationist strand in the environmental movement, while Marx provided a starting point for those who see the environmental crisis as the dying throes of capitalism and believe that the first stage of reform must be the overthrow of capitalism itself.*

Chapter Three

The parable of the talents

Reforms are likely to succeed or fail in the long run according to whether they are able to build on existing traditions, and for most of human history the moral and political traditions which have carried most weight have been derived from a religious context. The religious context in which the civilisations which now pose the most serious threat to the environment have developed is that of Christianity, which, like other religions, was the product of a particular relationship between society, in this case European society, and its environment.

This is to assert a particular theory of the relationship between culture, religion and ethics which should be made explicit as a basis of the following discussion. Society begins not with religious dogma but with codes of ethics which develop in human societies to save time in bringing up children. Parents place actions in unambiguous moral categories of right and wrong to provide children with a shorthand guide to how to behave, and from these arise generalised moral propositions like 'obedience to chiefs' or 'consideration for others', which rationalise the ground rules. From moral propositions arise religious dogmas, derived, as a rule, from a cultural understanding of history, and these provide authority for good behaviour as judged by a particular society. Society thus produces the religion it needs to authorise functioning in customary ways. The Tsembaga sanctify custom to organise the planting of crops and the raising of pigs. When customs and perceptions change and require religious sanction, this can also be accommodated if it is not done so hastily as to call into question the whole epistemological system on which religious authority is based. Westerners have hallowed in turn, and not without strain, such contending notions as papal supremacy, the divine right of kings, the rights of property, individual liberty and social equality.

Some of the best illustrations of the moral developments which have shaped Western society are biblical. The books of the Old Testament are the products of a patriarchal pastoral society which was slowly becoming, during the course of its religious development, a settled and largely agricultural society.

By the time the stories related in the Book of Genesis were composed, and began to provide the basis of a moral education, Hebrew society had already set about the task of transforming nature, using a wide range of crops for food, planting vineyards and domesticating such animals as goats. In Genesis, man *justifies* his action; he did not subdue the earth or multiply because Genesis told him to; the stories justified the state of society retrospectively and provided authority for the generations of the future. For them, to 'be fruitful and multiply and replenish the earth and subdue it, and have dominion over the fish of the sea and over the fowl of the air and over every living thing that moveth upon the earth' (Genesis 1:28) was to be obedient.

The Old Testament, however, generally places God, rather than man, at the centre of things and the gap between man and the rest of creation is still a small one. Psalm 104, for example, lists the manifold works of God as having been created with marked impartiality for the benefit of a number of different species: 'He causeth the grass to grow for the cattle, and herb for the service of man: that he may bring forth food out of the earth' (v. 14).

But 'The high hills are a refuge for the *wild* goats and the rocks for the conies' (v. 18). Lions, too, 'seek their meat from God' (v. 21), and as for the sea, 'there is that Leviathan, whom thou hast made to play therein' (v. 26). The prophet Job relayed to the Jewish people from the divine whirlwind the rhetoric of a God who was not only all powerful but impartial in the distribution of natural benefits:

Who hath divided . . . a way for the lightning of thunder; to cause it to rain on the earth, where no man is; on the wilderness where there is no man; to satisfy the desolate and waste ground; and to cause the bud of the tender herb to spring forth? (Job 38: 25-7)

The New Testament, on the other hand, places man at the centre of the universe, and Christians, though accepting Genesis as the story of creation, adopted the view that nature was made for man. Christ asks, 'Are not five sparrows sold for two farthings and not one of them is forgotten before God?' (Luke 12:6), though the point of the story is that God is really much more interested in mankind than he is in sparrows. 'Even the hairs of your head are numbered. Fear not therefore, ye are of more value than many sparrows' (Luke 12:7).

The anthropocentric basis of Christianity was made more explicit in the writing of St Paul, which represents a fusion of Hebrew thought with that of the Greek Stoics. Paul's writing is the product of a society which has been living in a land 'flowing with milk and honey' for some time, of an intensively cultivated landscape. Paul quotes the Old Testament Book of Deuteronomy: 'Doth God take care for oxen?', and he is asking a rhetorical question. 'Of course he doesn't', is the answer, or if he does, not in the same way. Paul goes on to say: 'For our sakes this is written, that he who ploweth should plow in hope', that is, in the distinctive human hope of an extra-terrestial immortality (Corinthians 9: v. 9–10).

Support for this belief was derived from the Stoics, who justified the relationship between Greek civilisation and the environment by the 'argument from design'. Man's ability to subdue nature was evidence of his creation for that purpose. Paul's commentary, informed by a cosmopolitan education, is a useful gloss on the gospels, which are far from clear on the relationship between man and nature, and on the often contradictory legacy of the Old Testament. Ancient Greek religion taught that man should not try to become master of the world. That was *hubris* — an attempt to become a god. But from the fifth century BC onwards, the period of the Greek enlightenment, the concept of *hubris* was displaced by the argument from design. Aristotle argues in *Politics* that nature makes nothing imperfect, or in vain; the test of its perfection is usefulness to man. 'Plants are created for animals, and animals for the sake of man', and this was the idea which the Stoics of Roman society developed further. Cicero's work, *Concerning the Nature of the Gods*, argues that the produce of the earth is designed only for those who can make use of it. Cicero conceded that 'some beasts may rob us of a small part of it' but it should not therefore be concluded that the satisfaction of animal appetite was the end for which the earth produced good things. Oxen, indeed, were clearly designed to haul the plough, as could be seen from the shape of their necks (Passmore 1974, p. 14).

The notion of the illegitimate robbery by nature of what God intended for the use of men continued as a strand of Western ideology as least as far as nineteenth- and twentieth-century Australia, where kangaroos were regarded as robbers of pasture, as though pasture had been intended for sheep and hence man's legitimate profit, a 'talent' which the Aborigines had left buried in the ground, thus proving themselves to be unworthy servants.

When Christianity eventually became the official religion of the Roman Empire, which ultimately enabled it to become the dominant ideology of the Western world, it thus contained the potentiality for two lines of thought. First, it could be argued that man had an absolute mandate to sub-due nature — he need have no moral qualms about it — but at the same time the folly of over-exploitation was self-evident and it was in man's own interests to accept regulations. To overcome this problem, Christianity contained a second line of thought, the idea of man as the steward of God, which was a much more flexible remedy than the tapus which provided the answer for many non-European societies. Genesis 2:15 provides the text for this concept: 'and the Lord God took the man and put him into the garden to dress it and to keep it'. To apply to it, that is, the principles of good environmental management. The 'dark ages' from the fifth to the eighth centuries were a period of minimal environmental damage, though the survival of civilisation may have been tenuous. The settled societies which developed in the Carolingian Empire and the infant monarchies of Britain in the eighth and ninth centuries were, however, much more capable of changing the environment than any of their predecessors.

From the eighth century onwards, the small ploughs which scratched the surface of the soil in the hands of subsistence farmers, pulled by single animals, began to be replaced by ploughs which dug deeply into the soil and turned it over. This technological change was accommodated by the collectivisation of cultivation under the manorial system as the peasantry contributed the labour of themselves and their animals to co-operative production and submitted to a reorganisation of land tenure.

Christianity provided the ideological, the educational and the administrative basis for a society now organised for economic growth. Counter-currents to this mainstream eventually added strength to it. Monasteries, like Rievaulx in Yorkshire, for example, were established in wild and desolate places to escape the folly and corruption of the civilised world, but Christianity soon provided an alternative impetus for industry and intelligence and the reinvestment of profits from co-operative labour which made the Cistercian order perhaps the most effective pioneer organisation in pre-industrial Europe. They built water mills, improved navigation on rivers, diverted their courses if need be, traded in wool and invested the profits in further colonisation and industry.

By the time of Domesday Book in 1086, only 20 per cent of England remained under its original forest cover, a figure which was to be reduced to 4 per cent by the early twentieth century. But the injunction to dress and to keep the garden of Eden was not forgotten. The surviving forest was increasingly managed, and often replanted, especially from the fifteenth century onwards, in order to keep up supplies of shipbuilding timber. Increased production and the accumulation of surplus in the hands of the Church and the nobility allowed for an increasing division of labour, specialisation and the development of craftsmanship. Men managed resources for their own advantage in ways which were to create unfortunate precedents in the nineteenth century. Rabbits were introduced into England in 1176, without anyone being asked to submit an environmental impact statement, so that the peasantry could be better fed. Coalmining began in the eleventh century, leading to smog problems in London. Wind and water power were by then in use all over Europe, but the most significant technological change of the late medieval period was probably the pendulum clock. It was the first artificially powered machine in the Western world, representing an idea of unforeseeable power. It also signified the acceptance of a new kind of thought, concerned with measurement and accuracy, the basis of the scientific method, and the ability to proceed by reason and experiment from the known to the unknown. Roger Bacon, who was a friar and a philosopher of the twelfth century, was able to write to a friend and tell him: 'I have not seen a flying machine, but I have met a man who has worked out the principle of the thing' (Brewer 1859, pp. 532–3).

In all this, the mainstream of religious thought provided a rather negative kind of justification, but it sufficed to enable scientific enquiry and economic growth to proceed without causing major conflicts of conscience. Nature, according to Anselm, St Augustine and Thomas Aquinas, was not sacred. The application of technology to productivity was therefore legitimate. Aquinas taught that man had a natural right over material things, and because of his reason and will, could use them for his benefit: 'the less perfect fall to the use of the more perfect. Thus plants use the earth for their sustenance, animals use plants, men use plants and animals' (Gilby 1964, vol. 8, p. 125).

This was a safe enough position while the economy and population of Europe maintained what seemed to be a steady state', that Utopia towards which environmental reformers now aspire. It was not really steady at all, but colonisation of marginal lands and the slow increase in the efficiency of production, keeping pace with slow and irregular growth in population, meant that an ideology which equated the social and political order with that of nature was widely accepted.

Further and more rapid change from the twelfth to the sixteenth centuries included the development of industrial economies in the Netherlands, northern Italy and London, and, with the seats of government, the cathedral and university towns, they provided centres for the accumulation of capital and an infant banking and moneylending industry, servicing a growing population.

The medieval ideal of a society governed by Christian principles was difficult to reconcile with economic growth. For Aquinas, the safety catch which prevented man's right to utilise the environment from becoming a right to abuse it was distributive justice:

Earthly goods which a man has from God are his, as far as the ownership of them goes; not so with the use of them, which he must not keep to himself, but share with others, who can be supported out of what he has over and above his own needs (Gilby 1963-75, vol. 34, p. 255).

This made an effective alliance between technology and capital accumulation pointless, so there was a shrinkage of the areas of conduct in which Christian principles were applied literally and the Bible was scrutinised for guidance, and for loopholes. 'He who takes usury goes to Hell', lamented an Italian banker, 'he who does not goes to the workhouse' (cited in Tawney 1938, p. 24). More useful was the parable of the talents (Matthew 25: 14-30) which, in contrast to the egalitarian teachings relating to the difficulties facing rich men seeking salvation (Matthew 19: 24), and concerning the folly of laying up treasure upon earth (Matthew 6: 19-20), equated capital accumulation with virtue and implied that the maintenance of a 'steady-state' economy was a wrongful neglect of opportunity. Both the Catholic and Protestant Churches accordingly modified their teachings in conformity with contemporary moral standards.

At an official level, Christianity has always taken a militant stance against the naturalism of the pagan societies of Europe which it replaced. The worship of wells, streams, trees and natural objects often associated with fertility and healing, was outlawed by an invading patriarchal church as idolatory. But as at other times in other places, there was a significant blending process. Local saint cults, and miraculous traditions, often involving females, provided a vehicle for a perpetuation of indigenous reverence. The line which the medieval Church drew between heresy and sanctity was always fine, and often difficult to locate. The Church was sufficiently wise to base much of its calendar on pre-existing pagan fertility festivals such as Easter and harvest time, and the liturgy was closely tied to the temporal rhythms of the northern hemisphere, thus harnessing much of the spiritual energy of its congregations (Spretnak and Capra 1985, p. 245).

In the 1980s, a new group of theologians exemplified by the Dominican Matthew Fox and Thomas Berry, both of California, have rescrutinised the scriptures and the religious writings of selected medieval personalities in order to discover a tradition of 'Creation spirituality' within the Church rather than outside it. Against the 'redemption

Christianity' exemplified by such proto-puritans as St Augustine, they propose an alternative tradition in which the Fall, the doctrine of Original Sin, and the need for redemption take a minor place, and they point to the works of such mystics as Hildegard of Bingen (1098–1179), Mechtild of Madgeburg (1210–80), Meister Eckhart (1260–1329) and Julian of Norwich (1342–1415) as evidence of an equally authentic and ancient Christian tradition which views creation as a continuing participatory process, and which provides a theological basis for a reconciliation between humanity and a misused planet.

Sabina Flanagan, author of a recent biography of Hildegard of Bingen (Flanagan 1988) disputes the authenticity of this interpretation of Hildegard's Christianity, at least, and argues that much as she might provide an inspiration for modern eco-feminists and creation spiritualists, her own theology was firmly founded in the fall/redemption mainstream. More research may well produce more evidence of the widespread influence of Creation spirituality in medieval society, but if so, it will still be difficult to argue that it can be philosophically reconciled with orthodoxy, either Catholic or Protestant. Ought does not equal is, and as Fox (1983, p. 28) himself says:

But when it comes to human concepts there are either/or choices that we must make ... The West has been travelling the fall/redemption path for centuries. We all know it, we all have it ingrained in our souls; we have given it 95 per cent of our energies in churches both Catholic and Protestant. And look where it has gotten us. Into sexism, militarism, racism, genocide against native peoples, biocide, consumerist capitalism, and violent communism. I believe it is time we chose another path.

He may well be right, but for Christianity it will be a new path.

Some modern eco-politicians regard this tradition as one worth reviving and building on, a means of attracting both some Christians and some feminists into the movement; but for two centuries the teaching of the Church that the only alternative source of spiritual power was the Devil, meant that any older beliefs or practices were bound to be victims of the infamous 'witch craze' which became a major preoccupation of the authorities as the expanding society of the early modern period began to draw sharper lines between 'man' and nature (Murray 1921: Thomas 1971; Trevor-Roper 1972; Larner 1981).

Tudor and Stuart preachers emphasised the anthropocentric aspects of Christianity more and more forcefully. Elizabethans were taught firmly that animals were made for man's convenience. It was with human advantage in mind that camels were allotted to Arabia, where water was scarce. Savage beasts were sent to live in deserts, where they did little harm. Horses' excrement had been made to smell sweet, as one London preacher told his congregation, because horses were often in the vicinity of man. Apes were made for man's mirth, and the louse was there to provide a powerful incentive to human cleanliness. The sufferings of animals in the service of man were a consequence of the Fall, not of man's cruelty, and there was little that could or should be done to alleviate it. A favourite text of this militaristic, imperialistic, hierarchical, bloodthirsty society was Genesis 9: 2–3:

The fear of you and the dread of you shall be on every beast of the earth, and upon every fowl of the air, and upon all that moveth upon the earth, and upon all the fishes of the sea; into your hand are they delivered. Every moving thing that liveth shall be meat for you.

It was a short step from the emphasis on the *permission* of God to utilise the environment to human purposes to the belief that technological advancement and success in moulding the environment to human purposes was the reward of Protestant virtue or evidence of God's approval. This was especially true among the evangelicals whose followers multiplied in the fertile soil of the early industrial revolution. The London Missionary Society resolved, on 28 September 1795, to introduce, as missionaries to the Pacific islands, 'not learned men, but godly men, such as understand mechanic arts' (cited by Newbury 1961, p. xxx) anxious, above all things, that the notion that Christianity went hand in hand with technology should not be lost on the heathen. Not surprisingly, the idea that 'the English God' had mana which enabled his followers to achieve the technological supremacy to which ships, firearms and implements bore witness proved a powerful inducement to accept the missionaries as guests.

Similarly, in the early twentieth century, Christian apologists were anxious to establish that the discoveries of science had the blessing of the Church, in spite of the anti-scriptural implications of some of them. Lynn White, President of the American Historical Association, on the occasion of his paper on 'The Historical Roots of Our Environmental Crisis' (1967) was more emphatic than most in attributing present-day ecological problems to the legacy of Genesis, and quoted Ronald Reagan, then merely Governor of California, as spokesman of the Christian tradition when he said, it is alleged, 'When you've seen one redwood tree, you've seen them all'.

It is true that Christendom produced St Francis, since 1981 patron saint of ecology, but as White points out, the wonder is that he was not burnt at the stake for heresy, as many good people were with ideas much less wild than his. The Franciscan order soon retreated, however, from the work of evangelism , ceased to work for social change and became a prop of orthodoxy. As such, the order could produce men like Roger Bacon, who thought about flying machines and introduced the experimental method into scientific enquiry, without feeling that orthodoxy was threatened.

There was, however, an inconsistency between the ideas of the Oxford philosophers of the twelfth century and the legacy of the Stoics, and it was to erupt in conflict at the trial of Galileo, who preferred the evidence of his senses and the fruits of the scientific method of enquiry to the authority of scripture. A corollary of the traditional teleological view was the idea that since God had designed everything for man's use, it was impious to tamper with it. From this notion arose the concept of the 'diabolical invention', a term applied to mechanical devices such as windmills, compasses and clocks, the fruits of a Faustian contract with the Devil. As late as the sixteenth century, a Spanish royal commission rejected a proposal to make two rivers navigable, on the grounds that man should not try to finish what God had left deliberately unfinished, a view which is unconsciously echoed in many modern conservation and preservation movements.

Authority for this view is derived from such passages in the teaching of Christ as the parable of the vineyard, which conflicts strongly with the usual interpretations of the parable of the talents, and it occurs in three of the gospels. The vineyard was planted by a householder, who leased it to a steward in his absence. When he sent his servants to collect the grapes, the steward beat them and killed them, so the householder sent his son, who was killed as well. Eventually the owner himself returned unexpectedly (the second coming), destroyed the wicked steward and replaced him with others who would hand over the grapes when due, who would not, in other words, exhaust non-renewable resources (Mark 12: 1–9; Luke 20: 9–16; Matthew 21: 33–41).

It is the concept of stewardship in Christian thought which perhaps has most to offer those in search of an ideology for a sustainable society, but a society which chooses to conserve resources for future generations cannot expect to rely on continuous growth to keep its masses contented. The corollary of the parable of the vineyard is the sixth-century teaching of Pope Gregory the Great: 'Our property is ours to distribute, but not ours to keep; we have no right of waste, and no right to withhold from those in need' (cited in Black 1970, pp. 63–4).

Secular ideology in Western Europe shares with religion a common classical ancestry, and the argument from design was the starting point for a developing intellectual response to changing technology. With the clock and the telescope, windmills, the mariners' compass, and now the New World incorporated into European experience, it was possible for Francis Bacon to believe in the sixteenth century that, through the acquisition of scientific knowledge, man could recover the position in the sight of God which he had lost through the sin of Adam. This, indeed, was man's purpose on earth: 'The end of our foundation is the knowledge of causes, and secret motions of things and the enlargement of the bounds of human empire to the effecting of all things possible' (Bacon 1627, p. 228). Descartes, on the other side of the English channel, argued the absolute right of man to do as he liked with the rest of creation. Animals were not only incapable of reason, he believed, they could not even feel, which was a rationalisation of the way in which animals were treated in early modern Europe. By the eighteenth century, the idea of a 'scale of being', including the whole of creation, was firmly established, with non-European peoples ranging from totally barbarous to 'semi-civilised', and slotted in above the animals but below Europeans, who were just below the angels.

These assumptions were not counteracted directly by any serious thinker until relatively recently, but they were much softened and modified by humanitarianism and the more charitable aspects of Christianity. Humanitarianism, as directed towards slaves or animals, was based on the premise that they shared with man the capacity to suffer. They were therefore, to that extent, human. This did not mean, however, that animals had acquired rights, but that man had lost them, and in particular the right, assumed by Descartes, that he could do as he liked with the rest of creation. New rights did not apply to animals which could not be seen to suffer in anything like a human way, such as ants or fish, and it clarified the distinction between plants and animals, which completely cleared man's primitive conscience, still found among the Tsembaga of New Guinea, and many other people, about cutting down trees.

J.S. Mill was the first political thinker to challenge seriously the assumption that man is entitled to the unlimited exploitation of the environment, and then only in a single chapter, 'On the Stationary State', which stands in strange contrast to the rest of the *Principles of Political Economy* in which it appears. He begins by arguing against Adam Smith, who had assumed that the condition of the mass of the people would always be pinched and starved in a 'stationary condition of wealth' and could only be satisfactory with continuous economic growth. Mill says that this may be true of the 'backward countries of the world', but that in the 'advanced' countries what is needed is better distribution of wealth. Having suggested what sounds like a left-wing solution to the problem, he goes on to say that the one indispensable means of achieving better distribution is a restraint on population, and he agreed with Thomas Malthus, who saw population growth as a guarantee that the majority of the population would remain in poverty, but was not overly concerned about the effect of population growth on the non-human environment. 'Levelling institutions, either of a just or an unjust kind, cannot alone accomplish it. They may lower the heights of society, but they cannot, of themselves, permanently raise the depths'.

Mill relates overpopulation not, as Malthus had done, to problems of food supply, but to man's psychological well-being, and defines overcrowding in a way which has a very modern ring about it: 'A population may be too crowded, though all be amply supplied with food and raiment. It is not good for a man to be kept, perforce, at all times in the presence of his own species'. He needs solitude, preferably in the presence of natural beauty and grandeur, 'the cradle of thoughts and aspirations which are not only good for the individual, but which society could ill do without' (Mill 1891, Bk 4, p. 497).

Mill is thus the starting point for two trains of thought, conservation of nature and populationism, which have formed an important part of the modern environmental movement, and, though it was not recognised to begin with, they can have extremely right-wing implications. William Vogt, writing in 1948, argued that the root cause of war was overpopulation, so a rational society should award sterilisation bonuses. 'Since such a bonus would appeal primarily to the world's shiftless, it would probably have a favourable selective influence.' Family benefits, on the other hand, though they had a levelling influence, supported 'hordes of offspring that both by genetic and social inheritance would tend to perpetuate their fecklessness' (Vogt 1949, pp. 282–3). He was the first of many to advocate that overseas aid should be given only to countries which agreed to control their population growth.

The Population Council and Planned Parenthood movement of the post-war years, financed by the Rockefeller Foundation, were fruits of this line of reasoning which developed an extremist wing in the late 1950s and early 1960s, just as the environment as a whole was becoming a matter of widespread concern. The issue which seemed at first to be central was pollution, and in industrial societies pollution was assumed to be the direct consequence of over-population rather than of particular industrial processes. Garret Hardin, in a famous essay, 'The Tragedy of the Commons' (1962), went further than even Ehrlich had been prepared to go and asked for immediate government action to terminate the freedom to breed. Other writers developed what has been called a policy of lifeboat selfishness: 'better that some should survive than none at all'. Aid should therefore be cut off to underdeveloped countries, and the affluent West should first bring its own population and resources into balance and then be prepared to defend its own against the inevitable military consequences of global recklessness.

Mill (1891, p. 496), however, also provides a starting point for a left-wing analysis of environmental degradation by criticising the assumption that affluence is a *sine qua non* of civilisation:

While minds are coarse, they require coarse stimuli, and let them have them. In the meantime, those who do not accept the present very early stage of human improvement as its ultimate type, may be excused for being comparatively indifferent to the kind of economical progress which excites the congratulations of ordinary politicians: the mere increase of production and accumulation.

Marx had no criticism of wealth as such, only of its manner of distribution, and the goal of ownership by labour of the means of production included ownership of natural resources and the means to exploit them for the benefit of man. A text which may have been missed by environmentalists is a speech he gave in April 1856 at the anniversary of the *People's Paper*, in which he anticipates the ideas of those modern environmental reformers who have criticised the qualitative consequences of industrialism.

In our days everything seems pregnant with its contrary. Machinery, gifted with the wonderful power of shortening and fructifying human labour, we behold starving and overworking it. The new-fangled sources of wealth, by some strange weird spell, are turned into sources of want. The victories of art seem bought by loss of character. At the same pace that mankind masters nature, man seems to become enslaved to other men or to his own infamy . . . All our invention and progress seem to result in endowing material forces with intellectual life, and in stultifying human life into a material force (Marx and Engels 1969, p. 500).

But for Marx there can be no going back. Industrial society is here to stay and its stultifying character is the price which must be paid for productivity. The question is simple; who is to control it? And the answer is predictable. It must be the Proletariat: 'We know that to work well the new-fangled forces of society —, they only want to be mastered by new-fangled men — and such are the working men' (*ibid.*).

Proletarian revolution thus has the answer to the qualitative as well as the quantitative ills of industrial society.

For modern Marxists, and the socialist wing of the environmental movement, it has been necessary to argue that socialist economies have inherited the environmental defects of big capitalism, size being the most obvious, and Marx has been carefully scrutinised to see if his ideas can be applied to a situation which he did not envisage. Malcolm Caldwell, for example, argues that the last stage of capitalism is the environmental crisis which results from overdevelopment. This is defined as 'taking more than a fair share of the world's scarce resources'. The flow of non-renewable resources is largely from poor countries, the victims of neo-colonialism, to metropolitan powers. What we are talking about is thus imperialism after all (Caldwell 1977, pp. 68-9).

From this he argues that there are specifically socialist approaches to the environmental crisis which are more likely to work than the 'palliatives pedalled by the liberal conservationists', but he does not become specific. Other writers, with similar views but less committed to Marxism as a panacea, include Herbert Marcuse and Ivan Illich, and though they would probably choose to be regarded as secular writers, they are difficult to place in a left–right political spectrum, and return to some of the assumptions of their religious predecessors. In contrast to the tradition of political thought, beginning with Hobbes and continuing through Locke, Descartes, Hume, Bentham, Marx and Mill, which is concerned with the environment only as it affects relationships between classes and individuals, Marcuse and Illich share with Aquinas an appreciation of the impact of man on the environment as a product of the relationships of human beings with each other.

In *Eros and Civilisation*, Marcuse agrees with the ordinary Marxist criticism of Freud, that Freud fails to see neurosis as evidence of a sick society, but, finding the defect in the individual, seeks to change the individual rather than the society. He goes on to argue that while some repression is necessary to any civilisation, most repression in existing societies, both capitalist and communist, is *surplus*, serving only to perpetuate existing inequalities and to keep existing elites in power.

One Dimensional man takes the argument further and links both inequality and neurosis with environmental degradation. 'The impoverishment of vast areas of the world is no longer due to the poverty of human and natural resources but to the way in which they are distributed and utilised'. Affluent goods, he argues, give satisfaction because people are manipulated into wanting them by a system in which profit depends on perpetual tantalisation. For Marcuse, the outlook for mankind is therefore bleak. The totalitarian tendencies of 'one-dimensional' society emasculate the ordinary methods of protest by allowing them to preserve the illusion of popular sovereignty but denying them effect. 'The people', therefore, previously the source of ideas and action making for social change, have 'moved up' to become the ballast of a conformist society addicted to growth. Much as discerning people can see disadvantages in the long run for this kind of society, no one can deny its popularity. It will be left to the substratum of outcasts in Marcuse's scenario to bring the present period of human history to an end, that is, when the young, the poor, the black, the female simply refuse to go on playing the game, but 'nothing indicates that it will be a good end'. Existing societies have the economic and technical capabilities to make illusory tactical concessions to the underdog while they all have 'armed forces sufficiently trained and equipped to take care of emergency situations' (Marcuse 1964, pp. 247–57).

Illich's (1973) diagnosis parallels that of Marcuse in many ways. 'Tools', that is, ways of making things, have become ends rather than means, and man has become an addict of growth. The more he has, the more he wants, and the more society must produce in order to avoid a revolutionary explosion. The withdrawal symptoms would be bloody.

But palliatives only make it worse. 'Consumer protection' is an example. People organise to get safer, better and more durable cars. This restores customer confidence in cars, and victory increases the dependence of society on high-powered vehicles, so more land is converted from food production and residence to highway construction. Planning, a favourite socialist palliative in Illich's view, is just as bad. So is the provision of social services: 'Once transportation, education or medicine is offered by a government free of cost, its use can be enforced by moral guardians'. The man who stays in bed to get rid of his flu instead of going to the doctor for antibiotics is sabotaging industrial growth.

So Illich predicts a crisis precipitated by a 1929-style stock exchange collapse, a depression or energy crisis. 'People will suddenly find obvious what is now evident only to a few: that the organisation of the entire economy towards the "better" life has become the major enemy of the "good life"'. As at other major historical crises, the Reformation, the French Revolution, the unthinkable can happen: 'people can discover that they can behead their rulers.'

But unlike Marcuse, Illich goes on to abandon metaphor and consider reality. The most probable outcome of contemporary development is, he believes, a revolution which results in a suppression of those aspirations which are now held at bay by tantalisation. A 'technofascist' regime replaces liberal democracy which addiction to growth, coupled with inequality, has destroyed. Illich regards this dead end as both politically and ethically unacceptable but prophesies its acceptance by majorities, at least to begin with: 'Such a *kakatopia* could maintain the industrial age at the highest endurable level of output' (Illich 1973, p. 101).

Some writers have shared the distaste of these prophets for the vulgarities of economic growth but they have been anxious to persuade environmental reformers that they should not be so disillusioned with the great accomplishments of Western philosophy and science that they abandon them completely and sit and wait for a revolutionary apocalypse. Fred Hirsch (1977) and Tabor Scitovsky (1976) have both sought to reform existing academic disciplines for the solution of the new kinds of problems which industrial growth has produced and they provide some hard evidence in support of the inspirational conclusions of the prophets. They may provide useful ammunition for the task of persuading majorities and governments.

Hirsch argues that once adequate food, shelter, plumbing, health and education are provided, together with a fair share of the conveniences of life, excessive growth becomes self-defeating. Once a certain level is reached, further growth, instead of raising everyone's access to the things they want, like jobs, country cottages, and so on, ensures further frustration. This is because extra growth makes things relatively less valuable, but harder to get.

Examples include suburban houses, which become less attractive as suburban sprawl spreads on and into the hills outside the city; private cars, which become a necessary liability when everyone has got one; doctorates, which cease to be guarantees of academic or any other kind of employment when increased access to higher education has made them too plentiful; seaside cottages, when there are so many of them that they spoil the view of the beach people buy them to enjoy. He concludes, therefore, that beyond the point which most affluent societies reached in the mid-1960s, growth is not a substitute for redistribution:

Capitalism has indeed brought the silk stockings that were the privilege of Queens to every factory girl; in this sphere and in this phase it has been a great leveller ... The achievement unfortunately does not enable it to repeat the performance with non-material aspects of privilege (Hirsch 1977, p. 188).

Tabor Scitovsky (1976) is another economist who is critical of some of the assumptions of orthodox economics. Its main weakness, he argues, is that it assumes that the consumer in the capitalist economy is rational, that whatever he does must be the best thing for him to do, given individual tastes, market opportunities and circumstances, otherwise he would not do it. This is the basis of the theory of 'revealed preference' which avoids the necessity to investigate any further the matter of why people do make the choices they make. Economics can then get on with the measuring and counting activities of an exact science which the public respects and on which businessmen, politicians and bureaucrats depend for advice on how to manage the national and global economies.

Scitovsky turns to the discipline of experimental psychology in order to test the validity of the theory. He finds that the rationality of consumers is not, after all, something that can be easily measured. First, there is the effect of advertising, which tends to undermine the parallel theory of consumer sovereignty. Then mass production and its economies of scale often provide choice only between different kinds of unlovely things. Most important is his demonstration of a clear separation between economic and non-economic satisfactions, the chief of which is work. Economists have assumed that the value of work is itself neutral; it is regarded for purposes of quantification as neither pleasant nor unpleasant. Scitovsky shows that it is chosen in many cases for the satisfaction it provides and avoided in others because it is boring or emotionally unrewarding, reasons which have nothing to do with its measurable money value.

Psychologists have discovered experimentally that in real life there is an imbalance between the twin poles of the Benthamite view of the world, pain and pleasure, which skews the results of attempts to test the felicific calculus. Once pain is removed from normal experience by civilised comfort, its absence is taken for granted and causes no particular pleasure. Pleasure, on the other hand, then depends on stimuli, for which people have an apparently insatiable appetite. The result is that people do not become happier as a result of the kinds of thing which can be most easily measured, like absolute income. Scitovsky relates public opinion polls about how happy people think they are to average income levels and shows that economic growth does not lead to correspondingly greater numbers of happy people. What makes some individuals happy is sometimes related to getting ahead of other people, but not always in terms of income or consumption; often it is in terms of such unmeasurable things as work satisfaction.

This is clearly illustrated by the way in which people with boring unfulfilling jobs have, in the last hundred years, used their bargaining power to reduce working hours while the managing director, who in the 1890s took a train to arrive in town at 10 a.m., took only a couple of hours for lunch at his club, and took the down train at 4 p.m., has been replaced by the sixty-hour-week 'executive' workaholic. The factory or office worker who used to work such hours a century ago now works only thirty-eight hours a week and saves living for that part of his life in which economists show little interest.

The ideological inheritance of the Western tradition is thus mixed, but not altogether discouraging from the point of view of the environment. For most of our recorded history, technological change resulting in increased exploitation of the environment by man has been justified by both religious and secular thinkers. But there have always been powerful counter-currents. It is not impossible that when faced by an urgent situation, a counter-current can be diverted into the mainstream by a sudden turn of events. Christianity itself was, after all, the sectarian belief of a minority in an underdeveloped colonial territory, but it became the orthodoxy of Western Europe because it was able to fill an ideological need.

Contradictory though the ideological legacy of the West may be, there is a consistent thread of concern for a sustainable and humane society which leads from parts of the gospels through Gregory the Great to St Francis and Thomas Aquinas. Submerged by the orthodoxy of early modern European society, it emerges in the humanitarian, liberal and socialist traditions of modern secular society, and in a Creation spirituality which may have a more ancient ancestry than the dominant forms of Christianity.

By the mid-1970s the environment had lost its initially fashionable position as a non-political issue on which all sensible people could agree. True, there was an emerging consensus on what was wrong with the relationship between man and the environment, but as the oil crisis, the war in Vietnam and the Watergate scandal dominated the international headlines, the dimensions of the environmental question began to seem impossibly enmeshed with everything else. The focus of debate began to shift from what was wrong to what should be done about it.

Suggestions for further reading

John Passmore's *Man's Responsibility for Nature* (1974) is both the best account and the best defence of the Christian tradition of stewardship and other kinds of traditional morality as a framework within which to address the environmental crisis. He argues that we do not need a new morality; we merely need to begin to comprehend fully the scope of our ethical inheritance and to take it seriously. Lynn White's presidential address to the American Historical Association, published in *Science*, on 'The Historical Roots of our Environmental Crisis' was one of the most important starting points of the modern environmental debate. For further detail, his *Medieval Technology and*

Social Change (1966), and Ffrancis Payne 'The Plough in Ancient Britain' (1947) are useful. Other historians who have contributed to the discussion are Keith Thomas in England, Hugh Stretton in Australia and Donald Worster in the United States. Carolyn Merchant, though concerned specifically with the historical origins of female repression in Western society, can also be seen as an exponent of 'environmental history', a new specialism. The 'witch craze', treated comprehensively in Max Marwick's *Witchcraft and Sorcery* (1970) has also attracted the interest of Hugh Trevor-Roper, of Christine Larner and of Keith Thomas, whose *Religion and the Decline of Magic* (1971) adds an important dimension to the literature on the birth of modern science.

Readers interested in the contribution of the thinkers of the eighteenth and nineteenth centuries to environmental philosophy will find R. Elliot and A. Gare, *Environmental Philosophy* (1983), a good introduction. Robert Attfield's *The Ethics of Environmental Concern* (1983) has more, and the collection *Ethics and the Environment* edited by Scherer and Attig (1983) contains excellent work by Sagoff, Goodpaster and Rolston. For the nineteenth century Maurice Dobb's *Theories of Value and Distribution since Adam Smith* is a good start.

There was, then, by the early 1970s, a substantial body of thought about the environment that provided a start for people who began thinking about it from a humanist point of view. But they were probably not in the majority. Many people who were concerned about the environment belonged to the other side of the divide between what C.P. Snow called the 'Two Cultures'. They set out on a path which originated in a particular area of scientific expertise, but it was eventually to lead them into the same intellectual tangle. Field naturalist societies, for example, often took the view that they had always been 'environmentalists' and were alarmed, when they attended meetings on 'conservation' issues, to discover crowds of unfamiliar faces. Marine biologists who actually measured the effects of oil pollution on mangrove swamps felt they had a right to speak and be consulted on environmental issues, since they had a proper understanding of what they were talking about. Plant pathologists smiled at good people who chained themselves to trees when workers came to chop them down to make way for motorways. They argued that people who talked about the environment when they knew nothing about the chemistry of a leaf did not deserve to be taken seriously.

It was the same with energy. Scientists saw this as a subject of concern to people who knew about it, like physicists. Geneticists were the people who knew about endangered species and so on. The damage which was laid at the door of science did nothing to dim its prestige, and the problems to which the public and governments had been alerted in the 1960s became the subject of one official enquiry after another, in both communist and non-communist countries, by committees headed by and mostly composed of scientists. No one thought this at all odd, least of all in universities. The fact that the first alarmists had been scientists added to the belief that what science had caused, a now all-powerful science, at the height of its prestige, would be able to cure.

But scientists were just as able as other people to learn from intellectual experience, and often more so. Whatever their particular scientific background, they were quick to learn that environmental problems, like dirt, were easier to shift around than to eliminate. One discipline after another came to be seen as relevant to environmental science, and in the period of academic expansion which coincided with what may prove to be the last ever period of universal growth, Centres of Environmental Science sprouted on university campuses all over the world. They were intended to provide governments and industry with broad-minded but expert analysts of environmental problems, proposers of cures and architects of policy.

Feedback from the 'real world' led to a further expansion of horizons which some people saw as a weakening of academic standards. Some Centres began calling on economists, lawyers, political scientists and philosophers to provide course components. What had begun as a cautious exercise in applied science became concerned as much with ethics as with enzymes.

The next chapter briefly explores the historic relationship between science and society in order to explain the structural difficulties of a purely scientific approach to environmental problems. Using a typical example, it shows how the search for scientific solutions to the problems produced by the relationship between science and society leads to the next discovery, that the problems which looked purely scientific were symptoms of deeper problems which were also political and moral.

Chapter Four

Science to the rescue

The belief that science has the answers to the problems which have been created by the misuse of science is not to be lightly dismissed, especially as it just may prove to be our only hope. Nor can it be denied that a development such as fusion power, now a real possibility in the foreseeable future, would go a long way to solving many of the more immediate environmental problems relating to the depletion of resources. Modern environmentalists would argue that such a breakthrough would merely accelerate the centralisation of political authority. It would increase the totalitarian aspects of industrial society, they would say, which would be required to contain the tensions of a necessarily violent culture. But one of the most powerful sets of arguments relating to the environment, as distinct from society, would be severely undermined. Ordinary people, with their power bills cut to a trivial level, would be much harder to mobilise in favour of environmental, non-nuclear or other peaceful causes.

Many people are concerned not about science itself, but merely the direction of much of modern scientific enquiry. If the effort now made to achieve the control of space, they say, were instead devoted to the production of synthetic protein, or the full development of the sea as a source of human nutrition, political and social imbalances might remain but the prospect of world starvation would be removed. Many writers have attacked the environmental movement on the grounds that it is opposed to man's unquenchable thirst for knowledge, and is therefore doomed to failure. It is not science as such which is to blame for the sometimes unfortunate consequences of scientific discovery, therefore an attempt to control or direct science in benign directions would be counter-productive and anti-intellectual. The most valuable products of pure research are often its spin-offs rather than its ostensible objectives. Therefore, it is argued, continuous pursuit of truth in a value-free context will ultimately produce the remedies as well as the problems. It is the politicians and the voters who must make intelligent use of the discoveries, which are neither good nor bad but inevitable (and so they should be).

Until the environmental crisis of the late 1960s these views were shared by the majority of educated people in industrialised society, but the sudden focus of public attention on what 'science' seemed to have been doing exposed scientists to a moral dilemma of ends and means which was the inevitable fruit of a period of increasing academic specialisation which had been going on since at least the thirteenth century.

For early medieval scholars there could be no moral dilemma since all learning was a means to the good end of the knowledge of God which provided a rationale for Christian ethics. Confident of the truth of revealed religion, and of the compatibility of all sources of truth, St Augustine's *De Doctrina Christiana*, written in the fourth century AD, was inspired by the conviction that all the sciences known to the pagan world had their place in a strictly Christian curriculum. Faith and reason, induction and dialectics were equally valid means of obtaining a view of the many-sided, and hence apparently differently shaped, mountain on which stood the City of God.

The twelfth-century recovery of the major works of Aristotle and the independent commentaries of the Arab philosophers strengthened rather than weakened the conviction of scientists that the tree of knowledge, which was good, could not bring forth evil fruit. Aquinas wrote in 1268 that 'the diversification of the sciences is brought about by the diversity of aspects under which things can be known' (Gilby 1963–75, vol. 1, p. 9). It was not until Robert Grosseteste, Adam Marsh and Roger Bacon, the Oxford philosophers of the thirteenth century, introduced the experimental method into scientific enquiry, that the first serious cracks in the philosophical edifice of Western scholarship appeared.

The right to pursue open-ended enquiry, well established by the sixteenth century, remains the foundation of academic freedom, and modern scientists have been concerned to emphasise their duty to pursue this goal regardless of social or political consequences. While, on one hand, they have maintained that scientific research must be funded from the public purse because of the technological advantages which flow from it, and because of the industrial applications of scientific discoveries, when science appears to be leading society down dangerous paths, as in the case of the environmental crisis, they have been quick to deny the idea of technological determinism. Science, it is claimed, is value-free, objective; the use which society chooses to make of scientific discovery is its own business. Technological predestination is denied by political free will. George Lundberg, for example, writing in 1929, claimed that: 'It is not the business of a chemist who invents a high explosive to be influenced in his task by considerations as to whether his product will be used to blow up cathedrals or to build tunnels through mountains'. He would have been most surprised if anyone had had anything against building tunnels (Lundberg *et al.* 1929, pp. 404–5).

As the effects of science on society escalated, so the disclaimers of moral responsibility grew more explicit. C.P. Snow (1965, p. 32) made it clear that, for his generation, innocence was an important dimension of academic free- dom. 'We prided ourselves that the science we were doing could not, in any conceivable circumstances, have any prac- tical use'. The more firmly you made that claim, the more superior you felt. The period between the wars seems in retrospect an Indian summer or moral irresponsibility. R.K. Merton pointed out, in 1938, the very real dangers of a science which could claim to be useful, in the light of German experience. In that country, he reported, 'Scientific work which promises direct practical benefit to the Nazi party or the Third Reich is to be fostered above all, and research funds are to be allocated in accordance with this policy'. In the same year, the Rector of Heidelberg Univer- sity said, 'the question of the scientific significance of any knowledge is of quite secondary importance when compared with its utility' (Merton 1973, p. 256).

The demands of totalitarianism made explicit the implications which scientists in democratic countries sought to avoid. Pure science had never been able to escape the stigma of its applied ancestry, in the workshops, arsenals and apothecaries of pre-industrial society. Now the threat of total war came to unmask its gentility. Sziland, a nuclear scientist, had a brainwave, so the story goes, as he stood at a red traffic light in the London Strand. On 2 August 1939, he wrote a letter, which Albert Einstein signed, to President Roosevelt, pointing out that, by invading Czechoslovakia, Germany had gained access to a supply of uranium and was on the road to making an atomic bomb, and suggesting that perhaps the US government should respond by channelling funds to speed up nuclear research in America.

British scientists were perhaps especially aware of the sinister implications of politically subservient science. The problem was to reconcile social responsibility with academic freedom; the movement for 'social responsibility in science' which was extremely vigorous in the 1930s, therefore took a negative rather than a positive form. Scientists objected to links which were increasingly established between science and warfare. The language of scientific discourse became deliberately impersonal. Scientists stopped saying 'I noticed that', or even confessing moments of personal inspiration as they stood at traffic lights. Instead they said 'It was noticed that', implicitly denying the relevance of time, place, social context and personal responsibility.

In this period, the scientific dilemma became acute. Scientific authority rested to a large extent on the demonstrable power of science to do good things, like finding new cures for disease, and as science became more incomprehensible to the layman, so direct demonstration of utility became a more important condition for the continued social support of science, which became essential as science became increasingly expensive. But in the alarming context of totalitarianism it was not surprising that scientific disclaimers of moral responsibility, for good or evil, should grow more explicit. The distinction between 'pure' (virtuous) science and 'applied' (bad) science grew more vehement, epitomised by the boast of the mathematician, G.H. Hardy (1940, p. 90) that he had never done anything 'useful' and by the apocryphal high-table toast at a Cambridge college dinner: 'To pure mathematics: may it never be of use to anyone'.

High-table conversation was all very well, but the horse had bolted well before then. In 1919, a British Parliamentary Committee on scientific and industrial research urged the view that, while it would be positively dangerous for the government to seek to direct research which it funded, science had been of great value to the war effort and should be assisted because it would play a major part in post-war reconstruction (Curzon 1919). As science became still more expensive, following the Second World War, its political subservience even in democratic countries became more inevitable. In 1971 the Rothschild Report in Britain presented a 'framework for government research and development' which typifies the accepted relationship between science and government in all developed economies: 'This report is based on the principle that applied research and development must be done on a customer–contractor basis. The customer says what he wants, the contractor does it (if he can), and the customer pays' (Rothschild 1971, p. 3) This formalised the marriage between science and society which had been a *de facto* relationship since 1914. Affluence now became the new justification for scientific research. The more research was done, it was believed, the better off would humanity become — healthier, richer, wiser, better able to utilise natural resources to its advantage. The ability of science to deliver the goods demanded by total war and industrial society was seen as the reward of virtue. Bronowski (1964, p. 86) wrote:

As a set of discoveries and devices, science has mastered nature, but it has been able to do so only because its values, which derive from its method, have formed those who practise it into a living, stable and incorruptible society.

So in spite of considerable misgivings about the ruins of Hiroshima and Nagasaki, the ashes of Belsen, and a strong critical current within the professional scientific community, most laymen felt secure in the belief that, having placed their trust and their hopes for the future in science and scientists, they were in good hands.

Scientists responded warmly. Such was their confidence and their willingness to accept responsibility that when the environment became a public issue in Britain and the United States, it was seen as a purely scientific problem. Science may have been misapplied in some cases, through technical inexperience, but that could be overcome as more experience was gained and shared. Where misapplication was due to such things as cultural factors, then the answer was to address each problem as it arose in an objective scientific manner. Energy had now surfaced as the major environmental problem next to pollution, but that was seen as a matter of finding the best short-term technological solutions, nuclear power being the most obvious, to buy time until the inevitable big breakthrough in the use of fusion power. Cheap and limitless energy would then release humanity finally from the bonds of natural restraint.

The belief that in science, as in commerce, the pursuit of individual academic profit acted as an 'unseen hand' to the advantage of society, continued to provide comfort to scientists who had inherited G.H. Hardy's belief in the political neutrality of their activity. They could no longer deny the 'usefulness' of science, but they were equally adamant in their belief that the direction of scientific enquiry was determined by purely academic considerations.

Others have argued, however, that as it costs more and more to do most kinds of science, it is inevitable that the direction of scientific research and the uses to which scientific discoveries are put are determined by the most powerful groups in society, whether the party hierarchy in the Soviet Union and China, or the 'military-industrial complex' in the Western democracies. Brian Martin (1981), for example, provides a case study, that of research on the effects of supersonic air travel on the ozone layer. Two research teams setting out to study the subject came up with diametrically opposite conclusions. One of them 'proved' that supersonic transports did irreparable damage, the other 'proved' that they did not. When funding was investigated, it was found that the team finding in favour of the supersonic transports was funded primarily by the aerospace industry and the military; only those scientists working within universities were able to 'prove' that the ozone would be irreparably harmed by the continued use of supersonic transports.

Similar conclusions could be reached about the scientific examination of any number of contentious issues such as the safe disposal of uranium waste, whether it is 'acid-rain' or just car exhaust fumes that causes the death of German forests, or whether polar ice-caps will melt if we do not reduce our output of 'greenhouse' gases and stop cutting down rainforests. Such issues tend to be settled one way or the other, like any other kind of argument, by a dialectical process, counter-argument, more work on both sides, the weight of evidence in the light of common sense, and gradually changing circumstances.

But even if it is accepted that science is objective, the questions which scientists ask are inevitably determined to a large extent by the organisations which employ them. In Europe, the Soviet Union and North America the defence and aerospace industries provide the largest opportunities for graduate scientists, and at the present time about 50 per cent of the world's scientists work at defence science. The direct benefits to humanity are incidental rather than deliberate.

Funding by governments through grants to universities and national research institutions, and by corporate bodies through industrial fellowships, can be credited with at least some influence on the direction of research. It is not necessary, however, to subscribe to a theory that scientists are politically subservient to their paymasters to discern in scientific activity a tendency to social and political conservatism; that is, to accept the social and political status quo of a given society, be it the Soviet Union, South Africa or the United States, as a parameter of scientific enquiry. There are at least three reasons for this, which have nothing to do with the political allegiances of individual scientists, who, as a professional group, contain as great a political diversity as any other, and more diversity than many. The reasons are these: the educational methodology of science; the intellectual effects of specialisation; and the *industrialisation* of the scientific enterprise.

To deal with these reasons in order. In science, textbooks are the main means of instruction and, characteristically, one textbook lasts a whole generation with annual revisions. This easily creates the impression in the mind of the student of a 'whig interpretation of science', analogous to the 'whig interpretation of history' in which the past is seen as a process of continuous improvement, with new discoveries being built on old discoveries, rather than by the human reality of controversy in a historical context involving personalities, allegiances, proper competition for funds, power and prestige. The 'whig' view of science presents the opportunity to do original work as the reward at the end of a long and rigorous apprenticeship of conformity, and it is this which gives science its elitist image, in the same way as the discipline of having to learn Latin used to give students of law, medicine and arts an elitist image which originally relegated science to the newer tertiary institutions in the educational system.

Since 1950, science has become organised in an increasingly hierarchical fashion, which has the effect of socially conditioning those students who aspire to a scientific career. Research is normally done by teams which are led by those who have been most successful in getting funding for their research in the past, a process described by R.K. Merton as the 'Matthew effect' after the passage in the parable of the talents which explains: 'to him that hath much shall much be given, and to him that hath not, shall be taken away even that which he hath'. Team leaders in turn gain most of the credit for the joint publications which are the usual result.

The second factor, specialisation, is primarily, of course, the result of the expansion of scientific knowledge, but it is partly the result of competition between researchers within existing disciplines, and this has an important side-effect. To make a successful career, a research scientist must produce something original. This can be done with most certainty by pursuing research into ever smaller subdivisions of specialised fields. The aim, except in an industrial or military context, becomes not to produce knowledge relevant to a specific issue, but primarily to prove the capability of the researcher as a value-free and objective investigator.

Funding of 'pure' research of this kind is justified politically, however, because of the incidental spin-offs, and the unexpected but prodigious implications of some pure research for industrial, medical or military uses. No one expected, after all, that research into molecular structure in the 1920s would lead to the plastic industry, with its benefits and problems, in the 1960s and 1970s.

One result of such innocent specialisation is that it can lead to an assumption that the physical laws derived from physics or chemistry must have their parallels in laws which explain social, economic and political events, including things like war, criminality and conflict. This can lead, from the point of view of the individual scientist, to a belief which excludes the necessity to question any of the factors which contribute to the human condition at any particular place or time, such as historical change, methods of production, inequality, religious belief or class relationships.

An example is the way in which research is undertaken into things like road accidents. These are presumed to be an engineering problem and, to some extent, a problem of physics and ballistics. Studies of coroners' reports on why people get killed and where they get hit, together with work using dummies, enables people to design safer cars and better roads and justify the wearing of safety harnesses and crash helmets. The result is that for the same number of road users, fewer of them will be killed. Relatively, that is an improvement.

But given the way modern cities have developed in response to the motor car, and the disincentives which exist for the use of safer and less polluting means of transport such as railways, buses and bicycles, and the importance of the car-manufacturing industry in the economy, more people will get killed because more people will be travelling by car. Research in engineering or physics is unlikely to suggest such remedies as the subsidisation of public transport, building more bicycle paths or decentralising industry. Such action might result in lower expenditure on insurance, medical fees, hospitals, legal fees and crash repairs. It would also improve the environment, but it is unlikely that any road accident research unit would recommend anything more dramatic than a series of palliatives, making existing patterns of road use and travel less intolerable. This is because, as scientists, the job of researchers is to present value-free results, not to make radical proposals for the reform of society.

In some cases, however, the social and political implications of the implementation of scientific remedies for environmental problems are inescapable. They are not always desirable.

The enthusiasm of neurobiologists for their scientific speciality has led, analogously, to the publication of articles recommending psycho-surgery for 5–10 per cent of the inhabitants of United States ghettos, who may otherwise be prone to violent and anti-authoritarian behaviour, and particularly for some well-read and sophisticated black 'trouble-makers' in United States prisons. The implicit assumption here is that aggression is the result of biological faults in the individuals concerned. Therefore, the notion that there may be something wrong with society, and that some men and women, unlike laboratory animals, think they have a good idea what it is, is irrelevant. There is therefore no point in changing society. Scientific knowledge can be used to modify the individual. The fact that people have, quite often, in the past, succeeded in changing social arrangements through political and military means is not considered a good precedent.

Similarly, in the nineteenth-century debate about race and 'intellectual capacity', the existing class structure at home and imperialism abroad were justified in the name of Social Darwinism. As imperialism fell into disrepute after the First World War little was heard of a scientific basis for racism. The defeat of Hitler's Germany in the Second World War discredited the theory further still, and 1951 marks the height of the racial truce, when a Unesco study announced an international scientific consensus that in terms of intelligence, all men were equal, and if they were not, it did not matter.

Since 1960, however, decolonisation has been seen to result in circumstances such as the rise of Idi Amin and the Vietnam war. In the same period, Jensen (1969) and Shockley (1971) in the USA and Eysenck (1971) in Britain have once again sought to depict troublesome people like the blacks in the cities of the USA or the Irish still living in Ireland, as an intellectually inferior species.

The point is not whether blacks or Irishmen are stupid — though it is natural to be suspicious when 'objective science' confirms long-standing prejudices — the point is that this kind of argument can be used to sanction social or political arrangements which place the Irish or the urban blacks at a disadvantage. The arguments can therefore be used to justify the use of the technology of repression to keep them that way.

A third reason why science tends to be inherently conservative, in spite of the left-wing or liberal views of many scientists, is the industrialisation of the scientific enterprise.

The only people working in science who can be compared to the lonely apothecary of the Renaissance, or the inspired investigator of the early twentieth century, driven by an insatiable curiosity, are those with their own laboratories or in charge of laboratories which are publicly funded. In both communist countries and the West, scientists now work, characteristically, in large hierarchically organised teams, in which there is an increasing division of labour. Each scientific worker is needed for the fragmented partial skills he or she possesses, and bound to a purpose only fully understood by the project director. This process of industrialisation complements the distinction between neutral or 'value-free' science and the non-neutral uses to which it can be put, in that it further separates the worker from his or her product.

The process expanded, science by science, starting with chemistry then moving on into physics and some kinds of biology, exemplified in the development of applied chemistry on industrial lines in the laboratories of many pathologists, where the specialist who takes credit, responsibility and most of the financial reward, sees neither the patients nor any of their bodily fluids, which are tested by people trained only to carry out the tests required and to record their results. Prizes go to an elite, usually male, often for discoveries which could not have been made without the ability on the part of members of the scientific labour force, often largely female, to appreciate the significance of unusual results.

The division of labour leads, as in industry, to great efficiency, but it makes flexibility very difficult and it is a context in which radical criticism of the use or direction of research cannot be expected to flourish.

Once a large investment has been made in a scientific enterprise designed with a particular purpose in mind, such as the production of weapons, it is difficult for that purpose to be changed. All the arguments brought forward by conservative industrialists in the late 1960s against making smaller cars, or by the management of the British firm Lucas Aerospace in 1976 against the production of socially useful products (McRobie 1982, p. 98) have their parallels in arguments against restructuring the scientific enterprise.

By the early 1970s, the environmental problems which had arisen in industrial society were seen to be of sufficient gravity to warrant just such a radical restructuring of the scientific enterprise. Barry Commoner was perhaps the best known and most renowned of those many scientists, whose expertise led them to an understanding of the impact of modern society on the environment, and whose humanity led them to the conclusion that the situation called for a supra-scientific solution. Many scientists all over the world shared his anxiety, and began to be interested in an inter-disciplinary approach. 'Interdisciplinary' soon became the buzz word of the faculty common rooms, but the respon-sibility was seen at first to lie with the physical and natural sciences rather than the social sciences. The consequences of this can be understood by making a brief excursion into the recent intellectual history of a typical university in the capital city of an affluent country with an unusually strong reformist tradition.

The Faculty of Agricultural Science was somewhat iso-lated from the rest of the university, which regarded it with a mixture of pride and envy because of its independent resources, small numbers of students, and the outstanding achievements in research which were the result. It was in a spirit of public responsibility that in 1970 the Faculty initiated a proposal for a course in Environmental Science. The perception of the course reflected the environmental preoccupations of the period, which were mainly concerned with pollution and the contamination of the food chain from agribusiness and industrial sources. Distributive and moral problems were not yet on the agenda. The proposed course was therefore to be provided only for graduates in agricultural science, architecture, engineering, medicine and science. The year of coursework was to be followed by a thesis, not on an interdisciplinary problem, but based on 'research appropriate to their undergraduate background in some aspect of environmental science'.

From the point of view of environmental management, scientific disciplines can be divided into sciences which have to do with production, like physics, economic geology, chemistry — and, as a conceptual oddball, neo-classical economics — and the impact sciences, including botany, zoology and most of the social sciences, in which scientists are concerned to measure the impact of the effects of production on the environment. Both require funding. The production sciences are the basic ones, the impact sciences the more complex, but it is the production sciences which attract the largest proportion of industrial fellowships and, because of the presumed objectivity and greater prestige of these disciplines, production scientists have been those most consulted about such problems as the energy crisis, pollution and more recently, the greenhouse effect. Attention has therefore been focused on how to enlarge energy supplies or how to develop alternatives to burning fossil fuels, rather than the more environmentally sane policies of reorganising patterns of production, distribution and consumption so as to reduce the use of energy whatever its source, and to distribute it more equally.

In this intellectual context, it was perhaps inevitable that the 'environmental crisis' was assumed in our typical university to be a fairly straightforward problem in applied science. The matter of devising an appropriate course was therefore referred to an inter-faculty committee composed of several scientists and an architect. The coursework part of the degree was designed to include the whole spectrum of sciences, and was documented in sufficient detail to indicate that, at this stage, the proposal was supported by both production and impact sciences throughout the university. This paralleled the contemporary lay assumption of 1970 that the environment was a non-political issue on which, unlike Vietnam, everyone of good will could unite. The document outlining the coursework syllabus concluded with 'political and legal aspects' as something of an afterthought. The choice of words is revealing: aspects of what? Of a problem assumed to be a little untidier than most, but of general concern to scientists rather than to anyone else.

At this stage, it seemed to the committee that 'legal, economic and political aspects of the course required more detailed elaboration'. This resulted in the production of an illuminating document, introducing a new course component entitled 'Understanding, Control and Change of Human Motivation'. It was to comprise one-tenth of the course. The guiding philosophy of the committee at this stage clearly included the assumption that there were proper remedies for the environmental crisis of which enthusiasts within each discipline were well aware. It was a problem to be tackled in a systematic and scientific way, like any other, by reducing it to its component parts and dealing with each part within the appropriate value-free disciplinary context. The bulk of the course was therefore to be taken up with enabling students to discern the problems related to particular disciplinary fields and to achieve competence in measuring them and in recommending the proper remedies. The obstacle to the implementation of the proper remedies was seen as the relatively minor problem of human history and the social foibles which it had produced.

The men of 1970 were optimists, however; they believed in know-how. The problem of persuading society to adopt the appropriate remedies, which the sciences could provide, for the solution of the environmental crisis, was perceived in terms of applying the appropriate technology. The first suggestion was thus a course component entitled 'General Principles of Health Engineering'. This was subdivided into three sections: mechanical health engineering; social political and biological engineering, and understanding, control and change of human motivation.

This was by no means an unusual approach to the problem of the environment and it is reflected in the first wave of modern books on the subject such as *Blueprint for Survival*. The idealists of the time were particularly impressed by the US space exploration programme as a model of procedure. A problem of infinite complexity had been solved by reductionist methodology. The necessary research had been parcelled out over a period of time and then finally recoordinated. The big step for mankind was a paradigm of the way in which problems could be solved by collaborative research effort. Perhaps the as yet intractable problem of man's abuse of the earth could be cracked by the same strategy.

Such proposals were made in universities all over the world in the early 1970s, and they bore fruit in a number of different ways, depending on the intellectual climate in which they developed and the forces of opposition with which they had to contend. Some of the strongest opposition came from the sciences, and from the best academic motives. Excellence (almost as powerful a word as 'interdisciplinary') could not be achieved except by rigorous research and specialisation. It would be impossible to maintain academic standards while attempting to generalise about such unmeasurable concepts as the 'quality of life'. There was nothing in recent scientific experience to suggest that success would be denied to those who continued to tread the narrow path of intellectual righteousness.

In the particular example I have chosen, it was a biological agricultural scientist of much vision, with the political views of a left-wing idealist, who took it upon himself to overcome the resistance of his colleagues. He wrote a memorandum explaining that:

The basic concept . . . to avoid producing the narrow specialist, is often contrary to the aim of those scientists who train postgraduate students. The general course is therefore designed on lines which may be unfamiliar to some scientists . . . the problems studied by the students will not be of the sort that can be solved by the application of a single discipline . . . The single discipline approach has been the cause of many of our environmental problems.

Soon, an economist, a lawyer and an historian were included in the planning committee, and late in 1971 the committee agreed, after long discussion 'to recommend . . . that the title of the committee be changed to the Interfaculty Committee in Environmental *Studies*'.

The change in title was significant. Similar centres in other universities throughout the world have usually changed the content and emphasis of environmental science, whether or not they have changed the name, but it is important that the meaning of the change has mostly been recognised. Environmental *science* was the product of a state of mind which proposed a range of technological remedies for a problem of human, and therefore historical, origin. It implied that though change had to take place, society could stay as it was, since the environmental crisis had presented itself in forms which were assumed to be individually susceptible to scientific analysis and cure. If nuclear waste was dangerous, ways could be found to make if safe; if there was increasing social violence, police could be trained to deal with it and more security staff must be paid to protect likely victims. Environmental *studies* implied the insight that both problems were symptomatic of an underlying problem which neither remedy would cure. It implied recognition of the need to understand the nature and cause of the disease before prescription, that is, for the causes of environmental problems to be understood in historical and social as well as scientific terms as a prerequisite for the prescription of the appropriate remedies.

It was now recognised that the remedies for environmental problems might be political as well as technological. Some people outside the academic establishment had always believed that they were philosophical or spiritual. Others were worried that any cure of a fundamental nature might be worse than the disease. There was agreement, however, that technological remedies were only capable of dealing with symptoms. Cures would be found only as a result of asking the political and moral questions of diagnosis.

Suggestions for further reading

C.P. Snow and W. Bronowski have been the most popular and effective promoters of the science culture and have done much to facilitate productive dialogue. Francis Crick was more aggressive, asserting in the last part of his series of lectures *Of Molecules and Men* (1966), that 'the old literary culture was dying, the new, scientific culture was bursting into life', and that 'University administrators should try to see that their Universities become centres for the propagation of the new culture, and not merely homes for propping up an ageing and dying one.'

This perception has been challenged by a large number of scientists as well as humanists. Notable among them was Alfred North Whitehead, whose *Science and the Modern World* first appeared in 1925. Another was Thomas Kuhn, whose *Structure of Scientific Revolutions* (1962) argued that the acceptability of scientific theories was partly a function of the prevailing scientific 'paradigm' at the time of its promotion. Hilary and Steven Rose produced the highly critical *Science and Society* in 1969, followed in 1976 by *Ideology of/in the Natural Sciences*, which includes work by political theorists such as Andre Gorz and Michael Cooley as well as scientists.

The work which most influenced the writing of Chapter

4 was probably R.W. Home's (1983) edition of the series of papers given at an international symposium at the University of Melbourne in 1979 on the place of 'History and Philosophy of Science' in the academic agenda. The papers by Hugh Stretton, Alan Musgrave and John Passmore were particularly relevant to the general title 'Science under Scrutiny'. The educational implications of reductionism for environmental education have been explored by Frank Fisher and S. Hoverman in their paper given to the 1988 Anzaas Conference since published (1989), as 'Environmental Science: Strivings towards a Science of Context.' The political details are discussed by John Young, Ken Dyer and Sandra Taylor in 'The Politics of Environmental Studies' (1989).

The process which began with the assumption that the environmental crisis was a scientific responsibility had led to the conclusion that it could not be solved by science alone. Centres of Environmental Studies which started as interdisciplinary centres of scientific research soon began to admit students from economics, law and the humanities. Each extension of the field widened the terms of reference within which the problem was considered. Not only did the competence of particular disciplines come into question, but even the tradition of objective, value-free investigation which some people believed to be a rationalisation of the very historic forces which had led to the situation of crisis. This line of reasoning led some to the conclusion that the priority for environmental reformers was to get rid of capitalism. Environmental activists found themselves likened to tomatoes — 'they start out green and turn red.'

But for many who started out at least partly red, and opposed the war in Vietnam as members of a broad-based peace movement in the United States and Australia, the effect of their success was to turn them green. The significance of the struggle to many who came to appreciate its complexities, watching it on television, experiencing it at first hand, or protesting against it in the streets, was not so much that it had been a victory for communism; rather, it had shown that superiority based on high technology was not infallible. A peasant army, morally supported by self-immolating Buddhist monks, and using bicycles and booby-traps against gunships, napalm and defoliants had been both ideologically and militarily victorious. Idealists turned from the moral dilemma of not wanting their own side to win to what seemed to be the simple issue of ecology which would lead to a victory for everyone. The lessons of defeat might at least be applied to the problems of creating a sustainable society.

This was the context in which F.E. Schumacher's book, Small is Beautiful, first appeared on the bookstalls in 1973. It was not so much an attack on capitalism as on industrialisation of whatever political complexion, and it was soon translated into several languages and published in paperback editions, which gained instant popularity with the post-Vietnam generation in Europe, and especially in North America and Australasia.

Schumacher eased troubled consciences by providing non-Marxist reasons for deploring capitalism and some non-capitalist reasons for deploring Marxism. Not all of them were new. The tradition of protest against the dehumanising effects of industrialisation goes back to the English Luddites. Jacques Ellul had made a comprehensive attack in La Tecnique (1956, published in English in 1964) on the psychological effects of industrial technology. Gandhi had criticised the industrialisation of the Indian economy as much for its cultural and spiritual insensitivity as for its economic effects. Schumacher introduced and developed similar ideas to a new generation in the wealthy countries, reinforcing the mood which rejected the materialism of the 1950s and 1960s.

Chapter 5 discusses the harmful effects of present kinds of industrial growth in both rich and poor countries. The 'small is beautiful' idea has been criticised as a new form of Luddism — the movement of machine-breaking protest during the early years of the industrial revolution in England. In fact, Schumacher was careful to explain that he was not opposed to technological advance, only to technological determinism. Luddism is therefore briefly re-examined in search of similar ideas and useful lessons of experience. A major objection to 'appropriate technology', the proposed alternative to large-scale centralised production, is that population growth does not permit such a luxury because it cannot be sufficiently productive. This argument is considered in the context of non-Western cultural values and changing ideas in the field of international aid and development, and the advantages which a greater degree of equality would have for the implementation of environmental reform.

Chapter Five

Small is beautiful, but can we afford it?

Schumacher's own title for his book was 'Economics as if People Mattered', which might never have made it a best seller. It was the inspiration of his publisher to call it *Small is Beautiful*, a direct challenge to the received wisdom of the age. It was translated into many languages within two years of its first appearance and young people picked it up from bookstalls all over the world.

In the first pages of the book the idea is introduced that the discipline of economics operates in a moral vacuum. The chief villain of the story is J.M. Keynes, apostle of growth, who had warned the world, at least half seriously, that economists could not yet (in 1930) afford the luxury of morality:

For at least another hundred years we must pretend to everyone that fair is foul and foul is fair; for foul is useful and fair is not: Avarice, usury and precaution must be our Gods for a little longer still. For only they can lead us out of the tunnel of economic necessity and into daylight' (Keynes 1930, cited in Schumacher 1974, p. 19).

Schumacher will stand some personal introduction because the intellectual changes encompassed by his life foreshadowed parallel but as yet incomplete changes in the thinking of more conventional economists. Schumacher had come to England from Germany as a Rhodes Scholar at New College, Oxford. He studied economics and became one of Keynes's most sincere admirers. He spent the first years of the Second World War in an internment camp and then as a farm labourer in Northamptonshire, but kept up his academic interests. He had been working at Oxford and later in New York on international currency arrangements, and sent a paper on the subject to Keynes at Cambridge. Keynes was impressed, but counselled against publication for the time being. Not long afterwards, in 1943, he declared himself indebted to 'writers of several different nationalities' when he published an article which contained ideas very similar to those of Schumacher. Shortly before his death, Keynes said: 'If my mantle is to fall on anyone, it could be on Otto Clarke or Fritz Schumacher. Otto Clarke can do anything with figures, but Schumacher can make them sing'. It was Schumacher who was asked by the *Times* newspaper to prepare an obituary for Keynes.

Under the influence of the Marxist Kurt Naumann, a fellow prisoner in the internment camp at Prees Heath, near Shrewsbury, Schumacher became a rationalist Marxist atheist. He believed with perhaps the majority of intellectuals of his generation that the concepts of absolute good and absolute evil were not analytically useful. The only useful questions to be asked in such terms were 'good for whom?', 'evil for whom?' As an economist in post-war Britain, he came down firmly on the side of state planning, large-scale state monopolies, mass production and standardisation (Wood 1984, p. 139).

Schumacher's road to Damascus was the undramatic daily journey of an office commuter which he began to make, when the war was over, from his new home in Caterham, Surrey, to the London office of the National Coal Board, to which he was Chief Economic Adviser. Seeking escape, by way of contrast, he began to read travel books, which led him to the mystics and philosophers of the East. Confronted with such a diametrically contrasting way of thinking, he was forced by his own intellectual honesty to consider it seriously, or not at all. In some anxiety he wrote to his parents:

All the conclusions I had come to have to be thought through again. And it is not only thinking that is influenced . . . I have the feeling that I will look back to my forty-first birthday as a turning point for the rest of my life (Wood 1984, p. 230).

In 1955 he was sent as an economic adviser, funded by the

United Nations, to Burma. His specific task was to evaluate
the advice which the Burmese government was receiving
from a team of American economists. Like Joseph Banks in
Tahiti and Margaret Mead in Samoa, he was culturally over-
whelmed. He wrote to his wife:

There is an innocence here which I have never seen before, the
exact contrary of what disquieted me in New York . . . Even some
of the Americans here say, 'How can we help them, when they are
much happier and nicer than we are ourselves?' (Wood 1984, p.
244).

Unsatisfied with the superficiality of such impressions,
Schumacher was unable to remain intellectually disengaged.
He had already begun to attend classes in Buddhist medita-
tion and philosophy in England. Now he entered a
monastery in Burma as a guest for a brief period before returning to public life. The Burmese government was not
impressed with his advice. It ran counter to the conventional wisdom of his contemporaries which regarded
customary values as major obstacles to the economic
development of 'backward', 'undeveloped', 'developing' or
'less developed' countries (LDCs) as poor countries were
successively labelled. He put his views on record however,
in a paper entitled 'Economics in a Buddhist Country'. In
this he argued that the science of economics lacked the
objectivity which it claimed. Far from being value-free, it
was derived from a particular view of the purpose and meaning of life, whether the individual economist was aware of
it or not: 'the only fully developed system of economic
thought that exists at present is derived from a purely
materialistic view of life' (Wood 1984, p. 247).

Since the modern science of economics had been
developed in a spiritual vacuum, what had been transferred
to countries which retained a spiritual tradition of their own
was not Western philosophy, but merely Western demands.
Western philosophy had made economic development possible, but it contained balancing potentialities for good
stewardship and the acceptance of responsibility as the
corollary of power. The transfer of Western demands
together with a failure to transplant Western philosophy as
a whole meant that the most popular things in poor countries were not what conscientious economic advisers wanted

people to have, but horror comics translated into Burmese, B-grade movies, alcohol, commercial sex and Coca Cola. 'The whole orient', Schumacher wrote to his wife, 'is coming out in western spots', (op. cit., p. 245).

Development in poor countries demanded action in accordance with indigenous philosophy, and here, lacking a 'fully developed system of economic thought', Schumacher drew on Gandhi, as well as on his immediate experience of Burma. Using the example of railway freight policy, he conceded that the demands of industrial society made sense of the practice of reducing freight tariffs for long hauls. This encouraged long-distance transport, promoted large-scale specialist production and led to the 'optimum use of resources'. But in doing so, it positively recommended one particular way of life. A Buddhist economist, having different purposes, would be anxious to promote a quite different way of life and would know that he was doing it. If he used subsidies he would support short-distance transport, and he might discourage long-distance transport in general, because it would promote urbanisation, specialisation and the growth of a rootless proletariat. The kind of economics preferred was thus inseparable from the philosophy and morality which inspired it.

It was therefore not to be expected that a materialist philosophy could underpin an economic view which appreciated the distinction between renewable and non-renewable resources. What mattered was only their relative cost at a particular time. Schumacher argued that the distinction should be considered in the same way as the distinction between income and capital. A civilisation which treated its 'natural capital' such as coal, oil and metals, as income, had long-term prospects no better than a business which sold off its capital assets to pay off its debts. Fifteen years later, Ehrlich, the Club of Rome and a lot of concerned scientists were to reach very similar conclusions without worrying themselves with the ethical considerations which had enabled Schumacher to get there first.

Small is Beautiful went on to draw further practical conclusions. Others were developed in the many lectures which Schumacher was invited to give all over the world, but most notably in the United States, in the 1970s. These were essentially reproduced in *Good Work*, published in

1979, while the philosophical base of his proposals, first developed as a course for external students at London University, was published as *A Guide for the Perplexed* in 1977.

Work was held to have not only a productive but a personal and social value. It was just as important as production and called for a different kind of technology. Like Gandhi, Schumacher distinguished between mass production, which required a massive injection of capital to create one workplace, and production by the masses, which created many workplaces with very little capital. He did not want to abandon the achievements of modern technology, but to see them evaluated and utilised within an ethical rather than a materialist framework. The best modern knowledge and experience would be the kind which succeeded in raising productivity by providing as many people as possible with both productive and satisfying jobs. It would also be conducive to decentralisation. It would make selective and efficient use of non-renewable resources, while preferring the use of renewable ones. It would enhance the pleasure and value of work to the worker rather than making workers subservient to machines. Mass production, by definition, was at best a solution to the problems of countries which were already rich and had the capital to fund expensive workplaces. In poor countries, productivity could best be raised and the problems of urbanisation and inequality best addressed by the adoption of 'intermediate technology'.

Buddhism provided a new global perspective to a tradition of social criticism which stretched back through William Morris, Ruskin and Dickens to the beginnings of that alliance between technology and capitalism which made the British industrial revolution at the beginning of the nineteenth century. Then, as in the modern industrial revolutions of the southern hemisphere, the social and environmental consequences depended on the ideological context in which industrialisation took place.

The British revolution in industry coincided with the intellectual acceptance, by the economic planners of the time, of Adam Smith's convenient notion that if each individual engaged in the uncoordinated pursuit of his own interests, he would ensure the best interests of all. This was the economic corollary of Newtonian rationalism, and the rejection of religious belief, except for formal purposes, by most intellectuals.

Lacking a clear sense of which changes were good and which were bad, Englishmen found the social and ecological effects of investing large amounts of capital and energy in industry very confusing. Political conservatives like William Pitt found common cause with those then seen as radical economists, Malthus, Ricardo and Adam Smith. Other conservatives, like John Doherty, the cotton spinners' leader, believed that 'if life were to be enriched by the new industry, machinery must be made subordinate to the interests of the men who used it'. They found themselves having to become revolutionaries in paradoxical defence of a traditional way of life.

Smith provided, together with much counter-balancing but now discarded wisdom, the central argument in favour of continuous growth, echoed by each generation of economists who have believed 'that the poor man in the rich community could live better than native kings'. In England this had the effect of encouraging Conservatives to dismantle the protective industrial legislation, inherited from the sixteenth century, which subordinated technology to the requirements of society.

The early industrial history of England has been likened to a civil war: 'The issue that now divided the English people was this . . . whether the mass of the English people were to lose the last vestige of initiative and choice in their daily lives' (Hammond and Hammond 1919, p. 4). Machines not only did thousands of people out of a job, at least in the short-run, but also, in the opinion of the displaced woolcroppers who, as Luddites, turned to violence and broke them, they deskilled the operators and replaced the creativity of the independent artisan with the mere productivity of the process worker:

The machines or *frames* . . . are not broken for being upon any new construction but in consequence of goods being wrought upon them which are of little worth, are deceptive to the eye, are disreputable to the trade, and therefore pregnant with the seeds of its destruction (*Nottingham Review*, 6 December 1811, cited in Thompson 1980, p. 581).

The Luddites were evidently less concerned than ecologists

would like them to have been with the use of power from non-renewable resources, or with the creation of the archetype of the polluted landscape That concern was left to the poets.

Historians of the right have tended to dismiss evidence of their pride in good workmanship as slightly spurious, and their motives as short-sighted, selfish or unrealistic. Those of the left have not been concerned with the ecological implications of the Luddite movement. The debate has focused on the much narrower issue of whether their aims were purely industrial, that is, in favour of maintaining high wages which were threatened by machinery, or whether the apparent industrial aims were a cover for a more serious political ambition, that of promoting a 'Jacobin' revolution at a time of war with revolutionary France.

What historians of the left fail to explain about the Luddites is why they believed that a revolutionary government would ensure the subordination of machinery to the values and purposes of society and in particular the maintenance of quality and creativity. Lack of sympathy by historians of the right is harder to understand, since the strategy of machine-breaking makes better sense as the last angry resort of conservatism than as serious industrial sabotage. Luddites never asked for new controls to make capitalism socially responsible. They objected rather to the abandonment by a supposedly paternalist government of what were seen as its existing, not its new, social responsibilities. They ended up on the scaffold instead of in the House of Commons because they failed to articulate their wider anxieties, of which their objection to particular machines was only a part. If they had been able to guess the historic significance of their predicament, they might have succeeded in gaining some powerful allies. They did not, and the British industrial revolution became a model for world-wide emulation.

Modern industrial society finds itself addicted to growth, at the expense of the environment, because of the problem which most concerned the Luddites — though it was Marx who found a name for it — 'alienation'. In the modern context, it is not alienation from the means of production in the sense Marx meant it, or even from the fruits of production, that really matters, but alienation from the process of production. The 'socialisation of the means of production' is all but irrelevant to the retention of a creative purpose if the process of production remains unchanged. Consider, for example, this parable of latter-day Luddism.

Management in a large transnational automobile factory in Australia was puzzled recently by repeated reports of faulty front axle action coming in from what seemed to be a random pattern of rural retail outlets. Computer analysis failed to come up with any meaningful findings. However, one day, an engineering student getting 'work experience' at the factory decided to leave the management offices and to take a walk onto the factory floor. He found that one worker was so bored by his allotted task on the assembly line that he took a heavy hammer at random intervals, and with it hit the front axle assemblies as they passed his bay (Peter Golding, pers. comm. 1984).

The environmental implications of such anger as this are the result of the place of frustrated workers in a circular causal chain of alienation, resentment and consumption. The jobs which large-scale manufacturing, processing, bureaucracy and administration provide for most workers have become increasingly repetitive. They are related increasingly to the servicing of machines which do the job rather than with the job itself. In such conditions the mind is less than fully engaged. The results include symptoms such as 'dropping out', absence from work, repetitive strain injury, accidents and hostility. These are partly symptoms of boring jobs, but they are also a result of the function of much of the work done in industrial society, which is to create the power to consume. The absence of a creative function undermines self-reliance and self-confidence, and makes industrial workers the easy target of an advertising industry which caters successfully for the fantasy life which makes it possible to endure reality. The intellectual and psychological vacuum created by the hardening of the distinction between work and leisure is therefore filled by consumerism. Wages are received not so much as payment for work, but as compensation for leisure denied, for it is only in the hours of leisure that the real business of living can begin.

The result is that the production of more consumer goods per head of population is a necessary condition of the survival of the industrial process, a conclusion reached by

the economic planners of the Soviet Union and China as
much as in Europe or America. Few people who have first-
hand experience of factory work or who are aware of the
dimensions of the global problems relating to energy,
population or pollution would disagree with Schumacher's
view that the technique of mass production, the hand-
maiden of consumerism, is 'inherently violent, ecologically
damaging, self defeating in terms of non-renewable resources
and stultifying to the person' (Schumacher 1974, p. 128).

There are, nevertheless, some reasons why a well-
intentioned ecologically minded dictator would do well to
consider the lessons of experience carefully before
implementing an 'appropriate technology' policy on either a
national or a global scale.

Within national boundaries, growth in the form of the
production of more consumer goods obscures conflicts of
interest so long as it is fairly rapid, and it is understandable
that many economists therefore regard increased growth as
the best guarantee of peace, both industrial and international. According to this argument, multinational and
transnational companies, which are good at causing
economic growth, should be encouraged. In the early years
of development, such companies acquired a bad reputation
as the agents of neo-colonialism — the control of the
economy of a poor country by a company whose major
interests lie in a rich one. Lessons have since been learned
on both sides. Representatives of poor countries, fortified by
the advice of young economists who have probably read
Small is Beautiful, are able to drive much harder bargains
with transnational companies than was formerly the case,
while their superior access to information, markets and
management skills can be made to serve national interests.
Relatively poor countries like Argentina now provide the
base for transnational companies which compete for
business with companies based in New York or London in
their own countries as well as in ex-colonial territories
which have a wide range of services to choose from. In the
long run the need for good customer relations is likely to
override the ephemeral advantages of a one-sided bargain
(Streeten and Lall 1976).

In practice, cynics might reply, it is the *possibility* of sharing the proceeds of growth which keeps majorities sufficiently

content to co-operate with the demands of the industrial economy. The factor which makes a widening gap between rich and poor tolerable in rich nations is that an expanding economy does allow for social mobility. In the 'twenty good years' between the wars in Korea and Vietnam, urban workers in Western democracies could rarely be persuaded to take an active part in class conflict. Most members of the working class could reasonably expect to leave the working class behind. They became 'upwardly mobile' and left the dirtiest and most monotonous jobs to the next generation to migrate from poor, hot countries to cooler rich ones, or from rural areas to the cities. They in turn would repeat the process faster still. It is not growth as such which takes the sting out of conflict, so much as the social mobility which growth facilitates.

There are nevertheless practical reasons for rejecting no growth as anything but a long-term solution to environmental problems, on the grounds that however desirable the ultimate goal might be in theory, the fights which would be caused by a sudden cessation of growth would be so rough as to make almost anything else preferable.

The results would vary with the ideology of the government responsible for making the decision. If, for example, a right-wing government decided that fossil fuels were really to be conserved for future generations, it would mean much more than just doing away with a few useless vulgarities. Using cars less when petrol was rationed in the Britain of 1940 meant that the car-owning minority had to accustom itself once more to walking and taking bus rides. Doing the same now, when cities have been built and public transport has been diluted on the premise of one or two cars per family, would mean that the people who would suffer most would be the poor, mainly women and children. They have been moved out of the old convenient inner areas near the city centres to make way for urban renewal schemes and 'prestige condominiums'. The renovated terraced houses are now occupied by affluent conservation-minded folk who can pay off their mortgages, enjoy the walk to work and lunch at vegetarian restaurants. The poor drive their rusting six-cylinder cars through peak traffic for up to two hours to reach their rented accommodation in the outer subdivisions (Stretton 1976, pp. 20–2).

One way to stop growth, and even to reduce unemployment at the same time, might be to work shorter hours. National governments tend to resist this in any case because if it is done unilaterally it makes it difficult for industries to compete with their overseas rivals. If agreement were possible it would not do much to reduce the use of energy or to decrease pollution. Domestic heat and air conditioning, leisure transport and recreation use up about 70 per cent of the energy used in a modern urban society. Less growth might even mean more work to produce the same amounts of essential things like food, warmth and shelter. It might mean harder work, for a longer time, using muscles and tools instead of computer printouts and machines. Majorities would be unlikely to prefer less growth yet.

The worst prospect of no growth or reduced growth is that it might *not* be accompanied by any redistribution of income, so that present inequalities would be carried over into a period of austerity, and the prospect of relief from poverty in the future would be removed. People can put up with almost anything under siege or during periods of wartime rationing because they look forward to the promise of peace, prosperity and plenty if they hold out. For the subjects of an ecological dictatorship there would be nothing in the future but more cold, austerity and hunger. Only with equality of sacrifice can self-denial be politically practicable.

The other, more common objection to a slowing of economic growth is that rising world population makes it impossible. Beckerman (1974, p. 239), one of the more optimistic economists in the limits to growth debate, argued that this makes growth the only option available. 'Since it is generally agreed that the progress now being made to reduce birth-rates cannot prevent the world population from doubling by about the end of the century there seems to be no alternative but to continue economic growth.' With the present annual increase of 70 million people a year, or two hundred thousand a day, this seems a very reasonable argument. If the present rate continues, the world will double its population in the next thirty-five years. Europe and Russia will double theirs in seventy-nine years, but Latin America will do it in twenty-four years. The argument for doing almost anything to achieve zero population growth as soon as possible is reinforced by the relative growth rates of urban populations which require large amounts of energy just to maintain the environment in working order. The quantitative approach therefore leads easily to the assumption of a direct causal connection between population and environmental problems. If one is reduced, it is argued, so will the other be.

These arguments, put so well by Paul Ehrlich that they convinced almost everybody, were a form of defence of industrial culture because they led to the conclusion that nothing else would be effective *unless* population was reduced first. Therefore, in the meantime, nothing else mattered. In fact, to reduce population growth in the parts of the world where it is most rapid would make very little difference to the environment. To stabilise population in India, or Burkina Faso would make almost no difference to the speed with which non-renewable resources are being used up. It would make very little difference to atmospheric pollution or the destruction of rainforest. What it probably would do is improve nutrition in those areas. If, on the other hand, you reduced business travel, air conditioning, space research, military expenditure, if you decentralised administration so that people commuted shorter distances, the pollution now caused by relatively small numbers of people in rich countries would be greatly reduced and they would use up irreplaceable resources much more slowly. A side-effect in some cases might be slower depletion of soils and a lower risk of starvation in ecologically vulnerable areas.

People constitute an environmental problem, not because of their existence, but because of what they do, and the parts of the environment they use up or damage. This is often related to their affluence. Generally, in affluent countries, the rich do most damage but the poor do quite a lot, too, because of the nature of the society in which they live and the extent to which they are victims of the cycle of alienation and consumption. In poor countries, the consequences of overpopulation are famine and overcrowding, rather than damage to the environment. This is not to say that starvation and overcrowding should be tolerated, but they are problems which are not likely to be solved by a simple reduction in population.

Birth control in poor countries has proved to be the

outstanding example of the limitations of the advertising industry. Slogans have notoriously backfired. The statement 'A planned family is a happy family', for example, is only likely to be true when it makes economic sense in a local context. When an extra child means an extra hand to chase pigs away from the vegetables, or to catch small fish for bait, that child will not be seen as a liability. That may come when schools are provided to which children must be sent, when old age pensions remove the need for children to care for the aged, and when good urban sewage systems reduce the infant mortality rate to the point where it is not necessary to have a large number of children to make sure that some, at least, survive. Birth-control campaigns have often been resented not for the economic theory behind them, but as an attempt to industrialise not only economic life but private life as well. Germain Greer (1985, pp. 405–6) succeeds, better than most, in making the imaginative leap from the air-conditioned office of the population planner to the urban shanties of Calcutta or Singapore or Durban or Rio:

It was the scourge of colonisation that cheapened human life . . . that showed the people in the hot lands that their destiny was not theirs to command. As long as the situation continues, as long as they have no resource base of their own, as long as they are mocked by the demands of foreign economies . . . they may wish to escape the pangs of childbirth, they may wish to escape the anguish of seeing children die, but they will not wish to be fewer.

It is misleading, moreover, to assume that economic growth, measured in terms of GNP, and starvation are alternatives. Much of the economic growth in poor countries since the 1960s has had little to do with food, and when it has, it has often been to do with the production of more nuts, coffee or meat for consumption in rich countries than with producing more yams or rice for consumption at the source of production. The best argument for economic growth in relation to overpopulation is that when its results have been well distributed and the living standards of the majority have been raised, an important side-effect has been the creation of conditions in which a large family is a burden rather than a means of survival. Better opportunities and some degree of social security have led to reduced growth of population. This is what has happened in Japan, Taiwan and South Korea. Where affluence is confined to an elite, as in the Philippines, the effect of affluence on population growth is reversed because people need their families to survive.

The problems which appear to arise simply from over-population were the major motivation for the creation of what has been called the 'development industry', the business of scholarly analysis and the provision of advice to both donors and recipients of international aid. Until recently, little attention was given to the ideas of Schumacher, who worked largely outside the academic establishment. Lucy Mair (1984, p. 6) argues that the sheer size of the problem of poverty makes the method of small-scale activity irrelevant:

Large-scale planners argue convincingly that the need to raise productivity in LDCs is far too urgent to be met by such tiny improvements, and that by themselves they could never meet the demands that are being made for a rapid advance to self-sustaining growth.

This misses the point of what Schumacher had to say. The changes he advocated were not primarily 'tiny improvements' of a mechanical nature such as windmills or bicycles, but ideas. The ideas, if consistently applied, in different circumstances were expected to lead to a wide range of technological choices which in some places might include windmills or bicycles. Paul Streeten (1981) has reviewed the progress made since the 1960s of the aid industry in an instructive essay which illustrates the parallel between the psychological development described by Schumacher's biographer and the historical changes of emphasis and purpose in the enterprise of international assistance at a rather later date, which took place as the lessons of experience in the field filtered through to the seminar rooms.

To begin with, the population 'explosion' impelled action in the light of the best analysis available in the 1950s. This was dominated by W.W. Rostow's (1960) doctrine of the stages of growth. Development was seen as a linear path along which all societies travel. The 'advanced' countries had, at various times, passed the stage of 'take-off', and the developing countries could be induced to follow them if obstacles such as traditional forms of land ownership, in some cases cultural or religious values, could be removed. Rich countries could help by coming in to 'break bottlenecks' by supplying missing components such as capital, foreign exchange, skills or management. Growth and inequality were seen as trade-offs. Inequality would have to be suffered to begin with so that capital accumulation would be encouraged. This would ensure growth of sufficient speed so that as wealth 'trickled down' to the poor, even they would be much better off.

It was not long before the theory came under attack. Some always doubted its moral and political wisdom and saw value in different styles of development compatible with different cultures. Others contended that the greater the gap between rich and poor countries the less relevant were the lessons to be learned from the experience of the rich.

Streeten considers Gunnar Myrdal (1957) the more scholarly critic of the linear view but Andre Gunder Frank (1966; 1967) the most influential, especially outside academic circles. Frank argued that linear development was in effect, if not in intention, a conspiracy between the elites of the donor and recipient countries to perpetuate inequality and poverty in poor countries. They favoured the kinds of development which increased urbanisation, created markets and sources of cheap labour for companies based in rich countries, and perpetuated dependence.

Frank's analysis became part of the conventional wisdom of the governments of poor countries and was generally accepted by the scholars of the aid industry, at least as a political factor in their calculations. There were two conclusions to be drawn. Aid might not be a temporary pump-priming exercise after all, but a permanent tax which rich countries must expect to continue to pay indefinitely to ensure international harmony. Alternatively, aid itself was part of the international system of exploitation and countries wishing to become self-reliant and independent should get rid of it.

Except for some countries like Uganda, which did not make a good precedent, dissociation from the wealthy countries was a political and practical impossibility, so the real question of the 1970s was how poor countries could pursue selective policies which would enable them to derive the benefits which were to be obtained from international aid while avoiding the disadvantages. The disadvantages were still real enough. There were plenty of ruthless transnational companies, some of them now based in relatively poor countries, offering short-term, socially disruptive opportunities to local profiteers for the sake of cheap labour. The World Bank entered its now discredited 'big dam' phase of development funding, with disastrous consequences for displaced rural communities and very unevenly distributed advantages for national economies (Goldsmith 1985). Educational investment could lead to forty thousand unemployed engineers as it did in India, or to a general brain drain, or to a refusal to consider the dirty jobs which still had to be done. But there were benefits to be had. There was a stock of technical, scientific and organisational knowledge available. There was money which was not tied to projects which would merely add to the markets of donor countries. It had now become evident to the aid scholars that the 'trickle-down' theory of wealth distribution did not work, and few poor countries had the ability to achieve redistribution democratically. Politicians themselves tended to come from the sections of society which stood to gain from established patterns of investment.

More recently, and in spite of much resistance, the objects of aid have been narrowed down to 'meeting basic human needs', specifically shelter, health, education, decent water supplies, sanitation and local transport. It is acknowledged that the linear development philosophy began at the wrong end, even though the big figures for *average* income, GNP and so on are often impressive. About one-third of the population of the LDCs still live below a nutritionally defined poverty line. There were 750 million of them in 1981. The consensus is now in favour of starting at the bottom rather than the top of the social pyramid and in the country rather than the city. Security of land tenure and access to local markets are now seen as the means to improve nutrition and rural productivity and provide an internal market for manufactured products.

There are no references as such in Streeten's survey to intermediate technology in spite of its obvious compatibility with the policy he now approves, but he does draw attention to the new emphasis on a recognition of interdependence among his colleagues:

The wheel has come full circle and we ... now acknowledge that many of the issues that we had seen as belonging to the poor countries are seen to be global, of concern to the rich too ... intermediate technology is just as relevant to high income societies suffering from unemployment in the face of resource limitations, pollution and alienation from work (Streeten 1981, p. 111).

Perhaps its relevance to poor countries can therefore be assumed. There seems to be a hint at least that the conclusions which Schumacher reached in 1973 about the best way of spending development money have been accepted:

Give a man a fish, as the saying goes, and you are helping him a little bit for a very short while; teach him the art of fishing and he can help himself all his life ... Teach him to make his own fishing tackle and you have helped him to become not only self supporting, but also self reliant and independent (Schumacher 1974, p. 165).

In poor countries the implementation of a 'first things first' aid policy will have an equalising social effect by making poor people better off. It will also reduce environmental damage by changing the direction of economic growth.

In wealthy countries, however, the effect of reducing growth in its present form may or may not be linked with more equality, depending on government policies. Encouraging labour-intensive rather than capital-intensive industry should meet opposition from nobody. It might even increase productivity, but such things as the practice of organic farming, craftsmanship and the installation of domestic solar heating remain the prerogatives of the economically secure. Getting less fortunate people to give up the lifestyle of consumerism will be difficult, partly for psychological reasons but also for financial ones. Old gas-guzzling cars are cheap to buy. People without ready cash can afford to buy cheap electric heaters in spite of the mounting electricity bills which will keep them poor. They cannot afford the installation of expensive slow-combustion stoves, or solar water heating systems.

One way of anticipating what might happen in a situation

of selectively pruned economic growth which was designed to achieve a change in its direction is to consider the situation of Britain during the Second World War. It is also an indication of what is politically possible in a democratic society, provided that two important conditions are fulfilled. First everyone must see the need for fairly drastic action; and we are getting close to that now in most developed countries. Second, it must be evident that there is a reasonable equality of sacrifice. The Second World War did a lot more to equalise British society, as it turned out, than the period of Labour government that followed it, and without that increase in equality, the necessary solidarity needed for survival would not have been achieved.

The purpose was to wage war effectively, but the side-effects were as valuable for the environment as they were for society. Apart from purely military activity, Europeans on both sides consumed fewer raw materials and polluted their atmosphere and their water less, as a corollary of greater equality. Good environmental habits became customary. People repaired things. Containers were all returnable. Newspaper was recycled. Scrap metal was ruthlessly recovered. People walked, rode bicycles and used buses and trains because petrol was rationed. Marginal urban land was used intensively as householders dug for victory and composting became a topic of polite conversation. The ration book in Britain proved itself what the revolver is supposed to have been in the United States, the great equaliser. It ended the situation in which social class coincided largely with physical size. In Britain more than half the population were better fed during the rationing period than they had ever been fed before. Wartime children were usually larger and healthier than their predecessors. This shows what might be possible in the way of environmental reform if the need were widely felt to be urgent, and if equality became the precondition of sacrifice.

Unemployment is often seen as a knockdown argument against any suggestion of slowing down growth. With current unemployment rates of around 10 per cent, how much worse would it be if there were less investment? Yet though we have continued to have growth, unemployment has also got worse, especially among school leavers. Unemployment could be seen instead as an opportunity for the application of Appropriate Technology, since the cost of creating each workplace would be greatly reduced. The social benefits might also be very great. George McRobie, author of a sequel to Schumacher's work, *Small is Possible* (1981), has turned his attention especially to the prescription of appropriate technology in advanced industrial societies. A text for this message is the activity of the Lucas Aerospace shop stewards, who began a movement in 1968 in response to a situation of declining orders for Lucas products from the space and military markets.

They mounted a campaign, not for increased wages or shorter hours, but for the right to work on socially useful products. Their literature emphasised the ethical abyss between what technology could provide, such as simple inexpensive heating systems for old people, enough dialysis machines to prevent the deaths of 3,000 kidney patients a year in Britain, and what technology does provide: the Concorde, exocet, space travel and so on. It pointed out that instead of freeing people from monotonous, repetitive work, computers, robots and automation generally have increased it and, like the steam loom before them done people out of jobs. The response of management was negative, but such initiatives create the opportunity for the redirection of production towards the creation of value rather than profit.

Such dichotomies may not be as real as they seem in any case, and a reforming government would defuse a great deal of hostility by using different methods of accounting. GNP is no more use as a measurement of human welfare in rich countries than it is in poor ones. Car manufacturing and crash repairing, for example, are both counted as aspects of growth, as are tobacco sales and funeral expenses, whereas one should really count as a cost of the other. Railways lose money and road haulage makes money. Both add to GNP, but having them do that does not show which kind of transport brings more benefits, or does more harm. Road transport has more accidents, leading to more deaths, more lawyers' fees, more days lost at work. It uses more fuel per ton per mile than rail transport. It requires more land for motorways than railways need, land which could be used for residence or production. It causes more pollution. These things should be taken into consideration as losses. Adding them to growth makes them seem like profits. Sailing cargo ships in terms of net human benefit (NHB) would come out well ahead.

Economies which look healthy in terms of GNP might look much less flourishing in terms of NHB if the violence, anomie, boredom and the diseases of affluence were set down on the debit side of the ledger to show that not all growth is good. Some LDCs, like Burma, Tibet and Western Samoa, would start to look as if they had some useful lessons for the rest of us.

Implementing such lessons would not mean less technological development, but its direction would have to be changed, especially now that the economies of several countries which have pinned their faith on yet another restoration of 'normal growth'; are getting into serious trouble. But such a general change in values is unlikely to be accepted without either a general catastrophe which forces it upon our collective consciousness, or a gradual change in philosophical orientation of the kind which took Schumacher most of his life and which is reflected in the aid industry since the 1960s. Without this, 'small is beautiful' is likely to remain a declaration of piety, but with some redistribution in the direction of equality, starting most easily and painlessly in the more equal rich countries, not all growth would be bad.

Suggestions for further reading

Schumacher had a number of precursors in the anarchist tradition, notably Peter Kropotkin, while William Morris, whose political writings have been collected and edited by A.L. Morton (1973) was concerned by the social impact of machinery. Kropotkin anticipated Schumacher's belief in 'good work', and also his belief in the virtues of a society based on small units. He imagined that village communities of the future would have up-to-date machinery in their communal factories, which would be small, and sited in close conjunction with the food-producing fields and gardens. His ideas are described in George Woodcock and Ivan Avakumovic, *The Anarchist Prince* (1950) and his own *Memoirs of a Revolutionist* (1899).

Apart from E.P. Thompson's famous work *The Making of*

the English Working Class (1980) the best work on the Luddites is by Malcolm Thomis. The modern and practical possibilities of creating work which is non-alienating and produces socially useful goods is discussed by George McRobie in *Small is Possible* (1981), by Mike Cooley, in 'Contradictions of Science and Technology in the Productive Process', his contribution to Hilary and Stephen Rose's 1976 collection of essays, and in his own *Architect or Bee?: The Human-Technology Relationship* (1980). Jack Mundey's contribution to Hutton's *Green Politics in Australia* (1987), is part of his developing strategy for bringing union power to bear on the problem.

Africa provides the worst examples of the ecological and social consequences of inappropriate development. For a Marxist analysis, the best starting point is Walter Rodney's *How Europe Underdeveloped Africa*. A green rather than red analysis of the same problems, especially in India and Latin America, has been presented in many issues of *The Ecologist*, especially since 1985, when a concerted attack on World Bank development policies was mounted in vol. 15. Edward Goldsmith's work with Nicholas Hildyard, *The Social and Environmental Effects of Large Dams* (1988) seems, happily, to have clinched the argument as to whether this kind of development is part of the cure or of the problem, and led to a new emphasis, by the World Bank, on the financing of environmental management and social impact assessment rather than inappropriate mega-projects.

The shift in thinking described so far in the minds of individuals and popular movements has been complimented by a shift of environmental issues from the periphery of the political agenda to the centre. In one country after another, in many different states of economic development, the 'green vote' has moved from being a minor irritant to becoming a constituency for which major parties and interests compete and none can afford to offend.

This opportunity creates new problems. The multinational media have raised the profile of the environment to saturation level, and political leaders like Bob Hawke, George Bush and Margaret Thatcher have been quick to seek to co-opt the green agenda with talk of a 'balance' between development and conservation, and well-chosen campaigns on behalf of attractive single issues like the Antarctic wilderness or the Daintree rainforest. At the same time, they continue to promise increased value-free economic growth in the context of a deregulated global market. The green movement is thus in danger of winning a number of very important battles, but of losing the war against the ideology of humanity versus nature and the materialism and consumerism to which that way of thinking has led.

If the political fruit of three decades of activity is to be harvested a degree of intellectual coherence becomes an important priority. The strategic need is for an ideology sufficient to unite an enormous range of single-issue organisations, including animal rights, zero population, Greenpeace, Friends of the Earth, the anti-nuclear movement, wilderness protection, appropriate technology, land rights for tribal peoples, feminism of different kinds, green socialism and conservationist conservatism.

The next chapter examines the major components of the constituency for their philosophical content and their potential for long-term compatibility. Hierarchical organisation and ideological rigidity are not necessary if there is a common purpose which can be clearly recognised, but, without such consensus, the history of socialism shows how idealism can be followed by disillusionment and, ultimately, the suppression of dissent.

The shift into the parliamentary arena has brought problems for both green parties and single-issue parties. Being 'neither right nor left but in front' is much easier in opposition than it would be in government. In government, the issue of social equity in a world now discovered to be finite becomes an inescapable corollary of democracy. As the sustainable societies of the past demonstrate, a common ethical denominator is an essential ingredient. It is something towards which most cultures can contribute.

Chapter Six

The ingredients of post environmentalism

By the beginning of the 1980s it seemed that the environmental movement had run into the doldrums. The sense of common purpose and optimism which had marked 'Earth Day' on the campuses of North American universities in 1970 seemed to have dissipated. There was plenty of tough new environmental legislation on the statute books, but it did nothing to challenge fundamental assumptions. Developers accepted that there would have to be a reasonable compromise between profits and conservation if 'sustainable growth' was to be achieved. They were quick to hire the new breed of environmental managers which the Centres of Environmental Science in universities all over the world were now producing. Those same centres and departments could not afford to be too radical, as they competed for diminishing funds with genetic and other kinds of engineers, computing scientists, and Schools of Business Administration. 'Natural Resource Management' became the favoured package for an 'environmentalism' of the 1970s which has been described as a search for means 'so that we can ravage the Earth with minimal effects on ourselves' (Bookchin 1987, p. 10). While, on the one hand, public support for measures aimed at 'cleaning up the environment' had grown by the beginning of 1988 in the USA to 80 per cent, less than 1 per cent considered the environment as the most important problem facing the nation. The editor of *The Amicus Journal*, Peter Borelli (1988, p. 39) took the view that 'Either the seriousness of the present situation has not sunk in, or else it has, and it's every man, woman and child for himself on a mad rush toward yuppie heaven'.

It was in the context of 'sustainable growth' that, in 1988, ministers of science and technology in a number of wealthy countries, sponsored a series of conferences on the 'greenhouse effect', the warming of the earth's atmosphere due largely to the emission of methane, carbon dioxide and other gases as a result of human activity. Simultaneous conventions, held in distant cities and hooked up to each other by satellite television, created an orchestrated and heightened sense of crisis. The aim was community involvement, but organisation was spearheaded by government agencies and keynote speeches were given by appropriate ministers and government representatives (Dendy 1988).

Agendas emphasised the need for hard data, the debatable nature of the phenomenon and the scientific doubts which remain about its cause. There was something of a hint, in Australia, that the whole exercise was an attempt to gain public support for a change in policy of the Labour Party government which would allow an extension of uranium mining and an expansion of the nuclear fuel industry.

If so, it backfired. The genie which the conferences let out of the bottle by encouraging popular participation was not, generally speaking, the technological optimism for which the promoters mostly hoped, but despair. Instead of discussing techniques for mitigating the effects of sea-level rises on coastal real estate values, workshops found themselves discussing suicide rates among teenagers and strategies for social empowerment. It was not only the relationship between the greenhouse effect and other kinds of damage to the physical environment which was obvious to a new generation of environmentalists, but the linkages between the damage which had continued to be inflicted on the earth, in spite of the rhetoric and the record of legislation of the past two decades, and the damage which the dominant ideology of economic growth was inflicting upon society (Fisher and Hoverman, 1989).

The hot summer drought in the northern hemisphere in 1988 was coupled with reports of dead seals in the North Sea, a proliferation of waste washed up on American beaches, ships roaming the seas like accursed fugitives searching in vain for some hapless poverty-stricken country on which to unload their cargoes of toxic waste, and media focus on the rapidity with which Brazilian rainforest was vanishing. The effect was to place the environment at the top of the international agenda once again. It was as if the spirit of 1970 was abroad once more, but in the main stream of public consciousness instead of the back eddies. Renowned champions of the environment like Paul Ehrlich, David Suzuki and David Bellamy found themselves addressing not just seminars of concerned scientists or committed activists but public galleries of thousands.

They were right to be apprehensive of becoming in the late 1980s what the popular evangelist Billy Graham had been to the 1960s because they were able to articulate and to tap the power of what has been recognised as 'a deep-seated innate spiritual concern for nature' which the secular society and the consumer culture have suppressed. It is this concern which provides the common ground for the diverse elements which may yet coalesce to comprise the 'post-environmentalism' of the closing decade of the twentieth century. Lorna Salzman (1990) regards the diversity of post-environmental thinking as a source of strength.

If the movement had demanded that its supporters joined for the 'right' reason . . . or that they hew to a single philosophy, it would have died a-borning. Instead it was able to make significant headway due to its inclusive appeal and its integrated strategies that fused science, politics and moral concerns.

This public and wide-reaching perception of the breadth and depth of the global crisis came to a head in 1988 and 1989, but its roots lay in many directions. One source of radical new ideas was, paradoxically, the NASA space programme, which may come to be seen as the last hurrah of the project of 'enlarging the bounds of human empire, to the effecting of all things possible' on which the scientific community was launched in the seventeenth century (Bacon 1627, p. 288). Professor Fred Hoyle anticipated as early as 1948 that, 'once a photograph of the Earth, taken from outside is available . . . a new idea as powerful as any other in history will be let loose' (cited in Borelli 1988).

That idea was the now famous 'Gaia hypothesis' originally proposed by James Lovelock at a meeting about the origins of life on earth held at Princeton, New Jersey, in 1969. It was developed at length in two major works, *Gaia: A New*

Look at Life on Earth (1979) and *The Ages of Gaia* (1988).

Lovelock was a member of the NASA team investigating the possibility of life on Mars and Venus. The paradoxical result of discovering that both planets were dead was the awesome realisation that the ability of the earth to support life was, in this solar system at least, unique. It was the novelist, William Golding, Lovelock's neighbour in the Devon village in which he lives, who suggested 'Gaia', the Greek earth goddess as a name for the hypothesis and, by implication, the planet to which it refers. Lovelock argues that the earth is a living being, but stops short of claiming that she is a spiritual being:

At some time early in the Earth's history before life existed, [that is, about 3½ billion years ago], the solid earth, the atmosphere and the oceans were still evolving by the laws of physics and chemistry alone. It was careering downhill, to the lifeless steady state of a planet almost at equilibrium.

This was in accordance with the second law of thermodynamics, which states that energy, once used, cannot be recovered and that natural processes always move towards an increase in entropy, a measure of the unavailability of energy.

But briefly, in its headlong flight through the ranges of chemical and physical states which were its destiny, a once-in-eternity, once-in-a-universe coincidence occurred, and the earth entered a stage which happened to be favourable for the existence of *life*, the paradoxical contradiction of the second law, in which evolution leads to ever greater complexity and diversity:

At some special time in that stage, the newly formed living cells grew until their presence so affected the Earth's environment as to halt the headlong drive towards equilibrium. At that instant the living things, the rocks, the air, and the oceans merged to form the new entity, Gaia. Just as when the sperm merges with the egg, new life was conceived (Lovelock 1988 p. 41).

Lovelock acknowledges that, like all important thinkers, he is a dwarf standing on the shoulders of giants, thus seeing farther than they (though few of them could write so well as he does). James Hutton, the 'father of geology' was first to put forward the theory that the earth is alive, in a lecture, given in 1785, to the Royal Society of Edinburgh. Hutton thought the most useful analogy for the student of the earth was physiology, and he compared Harvey's discovery of the circulation of the blood with the cycling of the elements, through the atmosphere, the soil, through plants and animals and back into the air.

Lovelock's evidence for his hypothesis is that in spite of the relative instability of the elements of the Earth's atmosphere, the fairly stringent set of conditions necessary for the continuation of life on Earth has been maintained. The proportions of the various gases in the different parts of the atmosphere have varied, as have the temperatures of the oceans and the extent of the ice-caps, but relatively, not by very much. Never in 3½ billion years have the oceans boiled or the ice-caps moved more than half way towards the equator from the poles. Lovelock argues that, just as our body temperatures remain fairly constant, because we shiver and sweat in reaction to extremes of outside temperature, and also because of the clothing we wear which has a function analagous to the clouds which enshroud a lot of the planet, so the living earth has maintained a similar constancy of temperature during a period in which the sun has grown 30 per cent hotter. This has been possible because once life was established, the biosphere continued to act as a self-regulating entity with the capacity to keep itself healthy by controlling the chemical and physical composition of the oceans, the atmosphere and the soil. The system thus includes the entire range of living matter, for example, the anaerobic bacteria which inhabit the mud in the bottom of ponds and produce methane gas. Without their activities the proportion of oxygen in the atmosphere would rise to the point at which any fire at all would get totally out of control and all vegetation would be destroyed. There would be nothing much to rot and the cycle of life would be broken.

At the other extreme in size, intelligence and complexity, but still within the system, are animals like whales, apes and humans, but from the Gaian perspective they are no more important. Living things do not evolve, it seems, according to fixed laws in a dead world, but in living circumstances which, as living things, they help shape. They may not always help themselves. Any species which makes the world uncomfortable to live in tends to encourage the evolution of species that can achieve a new and more comfortable environment. 'It follows that, if the world is made unfit by what we do, there is the probability of a change in regime to one that will be better for life, but not necessarily better for us' (Lovelock 1988, p. 177).

It is not the parts of the system, humans included, but the system as a whole, which is more than the sum of its parts, which has the overall capacity to maintain conditions for its overall health. Individual species may come and go.

Pollution, of such contemporary concern to humans, is thus a very relative concept. From the Gaian point of view it could be argued that the greatest pollution disaster in the history of the planet was the massive increase of oxygen in the atmosphere which killed off the surface-living ancestors of those anaerobic biota which now live only in the ooze at the bottom of the ocean and in our guts. Roses, on the other hand, thrive in the sulpherous atmosphere of London better than they do in the relatively 'cleaner' air of the countryside. The sulphur kills the fungus which attacks them (Lovelock, 1979, pp. 109–10).

The rhetorical device of indifference, which Lovelock consistently employs in both his books, was understandably disturbing to those of his scientific colleagues who inherited Francis Bacon's assumption that 'Man, if we look to final causes, may be regarded as the centre of the world, inasmuch that if man were taken away from the world, the rest would seem astray, without aim or purpose' (cited in Thomas 1983). It has taken the second environmental crisis of the 1980s for mainstream science to take the idea seriously. The results of a conference convened to consider it have now been published (Bunyard and Goldsmith 1988).

Would confirmation of the hypothesis mean that we are doomed by our biologically determined proclivities? Not necessarily, according to Lovelock, for these include the power of reason, and the ability to make choices. Gaia is not purposefully anti-human, but so long as we continue to change the global environment against her preferences, we encourage our replacement with a more 'environmentally seemly' species. It is therefore up to individuals to choose to act personally and politically in constructive ways.

Lovelock should not have been as surprised as he was when two-thirds of the people who wrote to him about the

Gaia hypothesis wanted to know what he thought about God. He seems ambivalent. On the one hand, he argues that the oldest level of religious experience is earth worship and the myth of the Great Mother, and credits the Indo-European civilisation which swept into Europe some 4000 years ago with the introduction of a remote master sky-god, as overseer of Gaia, legitimising in turn a warrior cult and a patriarchal social order. He asks

What if Mary is another name for Gaia? Then her capacity for virgin birth is no miracle. She is . . . conceivably a part of God. On earth she is the source of life everlasting and is alive now. She gave birth to humankind and we are part of her (Lovelock 1988, p. 206).

On the other hand, he draws back from the role of new messiah which he might have been tempted to claim. 'In no way do I see Gaia as a sentient being, a surrogate God. To me, Gaia is alive and part of the ineffable universe, and I am part of her' (p. 217).

There is nevertheless some consistency between the Gaia hypothesis and the 'minority tradition' of 'Creation spirituality' in Western Christianity to which the Californian Dominican, Matthew Fox, Wendell Berry, and others, have drawn attention (see above, p. 60). 'Creation spirituality' concentrates heavily on particular medieval figures in order to establish a traditional basis for its philosophy. This may not be necessary to establish its validity. There is plenty of evidence for the continuity of a religious sense of affinity for the earth in Western Europe outside the Catholic Church if not within it. Witness the Ranter, Jacob Bauthumely, who asserted that 'God is in all creatures, man and beast, fish and fowl and every green thing', the Romantic poet who claimed 'each shrub is sacred and each weed divine', and William Blake who believed that 'Everything that lives is holy' (Thomas 1983, p. 301).

More recent secular philosophers, such as Gregory Bateson (1980), have urged the utility of this attitude of mind in the interests of human survival:

If you put God outside and set him *vis-à-vis* his creation and if you have the idea that you are created in his image, you will logically and naturally see yourself as outside and against all the things around you. And as you arrogate all mind to yourself, you will see the world around you as mindless and therefore not entitled to moral or ethical consideration. The environment will seem to be yours to exploit. Your survival unit will be you and your folks or conspecifics against the environment of other social units, other races and the brutes and vegetables.

If this is your estimate of your relation to nature *and you have an advanced technology*, your likelihood of survival will be that of a snowball in Hell.

Aldo Leopold (1949) sought to respond to this dilemma, writing in 1948, by the development of a 'Land ethic' which would provide a consistent basis for an ecologically sound code of conduct for human society. He sought to enlarge the boundaries of that community to which ethical considera-tion was due, to include soils, water, plants and animals, collectively, 'the land', a concept which comes close to that of many indigenous peoples, such as the Fijian *vanua*, 'in short a Land ethic changes the role of *Homo sapiens* from conqueror of the land community to plain member and citizen of it. It implies respect for his fellow members and also respect for the community as such'. Lovelock (1979, p. 145) provides a scientific basis for such an ethic when he writes: 'The Gaia hypothesis implies that the stable state of our planet includes man as part of, or partner in, a very democratic entity'.

By the beginning of the 1980s the combination of scien-tific authority and philosophical disquiet had developed to a point at which a considerable part of the literate public in North America, Europe and Australasia was in search of a philosophical interpretation of nature as a rationale for social policy and human behaviour. Since then it has become, in the opinion of some, as important a focus of debate as Darwinian evolutionary theory was a hundred years ago, and nearly as divisive (Bookchin 1982, p. 214). This popular concern surfaced in the radical press, sand-wiched, to begin with, between writing on themes of more immediate concern such as nuclear disarmament and the women's movement, but the search for an ideological base became more intensive as campaigns by conservationists to save this and that became experiential exercises in radical philosophy. There is nothing quite like confronting a bulldozer with your legs chained to a buried log and only your head showing above the earth to concentrate the mind (Seed 1988, pp. 90–5). Winning frontline battles such as that over the Franklin dam in Tasmania focused attention on new intellectual problems. It meant, for example, talking publicly about the value of wilderness to tourists and the scientific value of rainforest as a source of medicine for human consumption. For many who did so it seemed that 'to save a wilderness, we had to betray its essence' (Fisher 1985, p. 17).

Murray Bookchin (1985, p. 216), one of the most outspoken and influential figures in the American green movement, warned that if academic philosophers ignored this rising theoretical trend within the social movements which they had discovered growing underneath their feet, they would merely add to the distance which already separated them from some of the most important developments in contemporary society.

First among academic philosophers to engage in the search for a new ethic was the Norwegian, Arne Naess (1973), whose earliest academic article in English on what he called 'deep ecology' appeared in 1973 in the journal *Inquiry*. Deep ecology was so named to indicate its basic premise, that the human species was no more than a plain member of the biotic community and that it sought to justify and rationalise a non-anthropocentric approach to living. Naess and his followers seek to describe the views of conventional environmentalists, however benign, as human-centred and therefore shallow: 'Thus even many of those who deal most directly with environmental issues continue to perpetuate, however unwittingly, the arrogant assumption that we humans are central to the cosmic drama, that essentially, the world was made for us' (Fox 1989).

The reaction of the philosophical profession was initially unfavourable, but this is now changing. John Passmore (1974) produced what is still the best defence of the dominant paradigm as an adequate framework for environmental reform, and dismissed deep ecology, along with its antecedents, in a lucid chapter entitled 'Clearing Away the Rubbish'. Richard Sylvan, Passmore's Canberra colleague, was still critical, but produced a generally sympathetic evaluation in the journal *Radical Philosophy* (Sylvan 1985). Professor Bill Devall and George Sessions of California published the totally sympathetic book *Deep Ecology* in 1985 and it became, in effect, the popular manifesto of a deep ecology movement. Meanwhile, new academic journals, *Environmental Ethics* (1979) and *The Trumpeter* (1983) began to publish papers by a number of Australian, Canadian and North American scholars seeking to iron out inconsistencies, to meet objections, and, by a dialectical process, to develop a coherent and intelligible philosophy.

A theme which has now become central is a redefinition of personal identity. Naess explains that 'self-realisation' is normally the term used to describe the competitive development of a person's talents and the pursuit of his or her specific and individual interests. It therefore conflicts potentially with other desiderata such as social bonds, friends, family, nation — and nature. Against this perception, Naess proposes the existence of the (comprehensive) Self — with a capital *S*. The maturing of a person consists in moving from self-realisation to Self-realisation, the latter consisting of identification with the whole of nature, so that the loss of a part of nature is suffered as a loss of Self. Thus, when Rachel Carson showed that DDT might inhibit the survival of Antarctic penguins, says Naess, people found it easy to make the first part of such a transition. It was easy to identify with penguins because they look so much like humans and seem to parody human frailty.

Thus ecology helped many to know more *about themselves*. We are living things. Penguins are too. We are all expressions of life. The fateful dependencies and interrelations brought to light, thanks to ecologists, made it easier for people to admit and even to cultivate their deep concern for nature, and to express their hostility towards the excesses of the economic growth societies (Naess 1985).

Deep ecology, Gaia theory and creation spirituality make an attractive but heady mixture. It may prove to be a less dangerous one than anthropocentrism mixed with advanced technology, but it is a good idea to know its political flashpoint. Anna Bramwell (1989) has pointed out that no sooner had the word 'ecology' been coined by the German biologist, Ernst Haeckel, in 1866, than it was made to serve as the basis of a radical social philosophy. Later, in the years following the First World War, it led to the foundation of the British 'blood and soil' movement, in which D.H. Lawrence was a prime mover. The now respectable Soil Association sprang from the same roots. Other offshoots grew more dangerously. The ecological idea appealed to the German Nazis because they, too, believed that the laws of nature could not be transcended by human society, and they opposed both capitalism and *laissez-faire* economics, the forerunner of today's 'economic rationalism', from ecological principles. They approved of organic farming, and Himmler established a number of experimental organic farms, including one at Dachau concentration camp. Hess was another disciple, who went in for biodynamic farming, homoeopathic medicine and naturism. His flight to England, however, tainted all such elements in Germany with treachery and from 1941 onwards the SS harassed both organic farmers and nudists, along with others who were seen as social deviants (Bramwell, 1989).

Others have pointed out the uncritical acceptance, by all deep ecologists so far, of the populationist arguments of Malthus, Ehrlich and Vogt. No distinction is made between rich, white and wealthy populations and poor, brown and black ones, the different kinds of environmental damage that they do, or the structure of the relationship between them which often makes ecological damage by poor populations the only way to survive. Deep ecologist Dave Foreman is quoted as saying, in an interview with Bill Devall in Australia, in 1987: 'When I tell people the worst thing we could do in Ethiopia is to give aid — the best thing would be to just let nature seek its own balance, to let the people there just starve — they think this is monstrous' (Bookchin 1987, pp. 13–14). Naess himself wrote, in 1986, more moderately:

It is recognised that there must be a long range humane reduction through mild but tenacious political and economic measures. This will make possible, as a result of increased habitat, population growth for thousands of species which are now constrained by human pressures (cited in Borelli 1988, p. 32).

The political implications are still alarming. Will majorities of people, especially poor people, be content to make way for more rabbits, and if majorities cannot be found to support such policies, will they be coerced? Who will decide when there are enough rabbits? It will not be the rabbits themselves. Deep ecologists have so far left themselves open too often to such attacks. How literally is 'biocentric democracy', for example, to be taken? Should we adopt a 'hands-off' policy towards such co-inhabitants of Gaia as the AIDS virus? Murray Bookchin, the most verbally violent critic of deep ecology, argues that historically a Self who absorbs all real existential selves has been used many times to absorb individual uniqueness and freedom into a supreme 'Individual' who heads the State, Churches of various sorts, adoring congregations and spellbound consti-tuencies. Ancient Egypt, Imperial China, Germany under Hitler, and Italy under Mussolini are cited as examples of the overenthusiastic application of the principle of the 'The Great Connected Whole' (Bookchin 1987, p. 19).

Deep ecologists are concerned to distance themselves from extremism of the Foreman variety, and have replied by argu-ing that the rights of humans in a biocentric democracy are not just no more than those of other species, but also no less. Conflicts of interest cannot be eliminated, and the vital interests, even of plants, involve the killing of at least some living things. Humans thus have the right of giving priority to their own defence, but they should distinguish between the defence of vital interests and the frivolous destruction of other species for the sake of mere advantage:

It may be of vital interest to a family of poisonous snakes to remain in a small area where small children play, but it is also of vital interest to the children and parents that there are no accidents. The priority rule of nearness makes it justifiable for the parents to remove the snakes (Naess 1985, p. 267).

Hunting cultures, which have a notoriously high level of identification with the hunted animals, kill the same animals for food. Naess argues that the function of ritual in such cultures is to express the gravity of the alienating inci-dent and to restore the identification. People living in rich countries, on the other hand, have no right to use the fur of endangered species for decorative clothing.

Lorna Salzman (1988, p. 6) argues that, though by eco-logical criteria humans are no more 'important' than other species, this does not mean that they cannot make moral choices. Like other species, they have the right to continue to evolve:

To make a decision that it is wrong to kill other living things for food may be 'moral', but it is counter adaptive to human survival. Any society which came to that conclusion would perish. Maybe some already did.

It is precisely our intelligence, she says, a part of our evolution that enables us to recognise the functional equality of species, which at the same time enables us to recognise our own right to protect our vital interests. Whales should therefore be protected, for they do humans no harm, but the elimination of smallpox virus is legitimate, because it does.

Though logically compatible with the Gaia hypothesis and with Creation spirituality at some levels, deep ecology has nothing to say about God, and is essentially atheistic. The proposition that the world was not made for the human species, or any other species for that matter, implies the absence of conscious purpose, and therefore of a God who might have had one. The basis of deep ecological ethics is therefore not the result of the development of a code of conduct, but the fullest possible identification with the cosmos.

Just as we do not need morals to make us breathe . . . [so] if your 'self' in the wide sense embraces another being, you need no moral exhortation to show care. You care for yourself without any moral pressure to do it (Fox 1989, pp. 34-5 quoting Naess).

Cosmologically based identity is something which Australian Aborigines have taken some fifty thousand years to develop and their cynicism towards those who claim to achieve this goal on a purely intellectual basis can be understood. This does not mean that it is not worth making the attempt.

The most outspoken critic of deep ecology in recent times has been Murray Bookchin, a man who can claim to have had a longer association with the green movement than almost anyone else. His first major work, *Our Synthetic Environment* (1962) was published under the pseudonym Lewis Herber, six months before Rachel Carson's *Silent Spring*, but, possibly because of its more politically radical message, it was not widely read. This was followed by *Post-Scarcity Anarchism* (1971), *Towards an Ecological Society* (1980), *The Ecology of Freedom* (1982) and *The Modern Crisis* (1986). In between books, some of which are collections of republished essays, there has been a constant stream of papers published in a widely dispersed range of journals.

His attitude to deep ecology is hostile, for he sees it as something which 'parachuted into our midst quite recently from the sun belts' bizarre mix of Hollywood and Disney-land' (Bookchin 1987). It is a hostility which seems unnecessary in view of the common ground he shares with his opponents. It is often the names he gives for ideas rather than the ideas themselves which are different. He believes that, on the one hand, there exists an ideology which he describes as 'environmentalism' and which he contrasts with 'ecology':

Environmentalism does not bring into question the underlying notion of the present society, that man must dominate nature; rather, it seeks to facilitate that domination by developing techniques for diminishing the hazards caused by domination. The very notion of domination is not called into question (Bookchin 1980, p. 59).

'Ecology', on the other hand, 'sees the balance and integrity of nature as an end in itself', a statement which deep ecologists would agree with. But the 'ecology movement' is, in Bookchin's view, deeply divided between 'a vague, formless, often contradictory ideology called "deep ecology" and a socially oriented body of ideas best termed "social ecology"' (Bookchin 1987, p. 13).

From the culturally distant perspective of Australia it does not seem that the differences are fundamental. The author of the following statement is difficult to identify as other than a deep ecologist:

Ecology knows no 'king of beasts'; all life forms have their place in a biosphere that becomes more diversified in the course of biological evolution. Each ecosystem must be seen as a unique totality of diversified life forms in its own right. Humans too, belong to the whole, but only as part of the whole.

But he is, in fact, Murray Bookchin (1980, p. 271). Differences centre around theories of causation. Bookchin asks rhetorically whether the cause of the ecological crisis is technology. No, people must choose sensible technologies and use them sensibly. Is it overpopulation, as the deep ecologists would have us believe? No, this is to make an effect of ecological dislocation into a cause. Is it mindless consumption? Here a half truth is used to create a whole lie, the 'original sin' of the 'unwashed consumer'. The poor cannot be blamed, as they are not responsible for the consumerist culture into which their economic masters have forced them, nor for the mirage of affluence for all, which so persistently retreats as soon as it seems within their grasp. No, the cause must be found in 'the underlying institutional moral and spiritual changes in human society that produced *hierarchy* and *domination* at the very dawn of civilisation' (Bookchin 1980, p. 39; emphasis added).

This does not seem so very far from the deep ecologist position in that it stresses the importance of moral and spiritual change as the necessary starting point for reform. Where it differs from deep ecology is in the assertion that the root of the trouble in which we find ourselves is the consequence of the domination of some humans by others; 'there is a tie in between the way people deal with each other as social beings, men with women, old with young, rich with poor, white with people of colour, first world with third, elites with masses — and the way they deal with nature' (Bookchin 1987, p. 12).

In one sense, this is merely to exemplify Stretton's 'environmentalist' observation that 'people can't change the ways they use resources unless they change the ways they relate to one another' (Stretton 1976, p. 3), but in the context of Bookchin's philosophy it amounts to a harnessing of ecology in the service not merely of egalitarianism, but towards the more particular purposes of anarchism. Bookchin (1980, p. 78) writes:

Voluntary simplicity will alter very little in our grotesque imbalance with nature if they leave the patriarchal family, the multinational corporation, the bureaucratic and centralised political structure and the prevailing technocratic rationality untouched . . . To ask powerless people to regain power over their lives is even more important than to add a complicated, often incomprehensible, and costly solar collector to their houses.

The rhetorical force of these assertions, and the considerable virtues of Anarchism, independent of its relevance to environmental issues, can easily disguise their logical limitations. There may be good reasons for regarding patriarchal families as obsolete, but they have existed for a very long time in a large number of cultures which have nevertheless achieved a high degree of 'harmony with nature'. Voluntary simplicity, if carried far enough, undermines the growth ethic on which the economic strategies of multinational corporations are based. That makes it a very self-empowering course of action. Compulsory simplicity would be especially unacceptable to anarchists, but it would be very good for the environment, as many countries have discovered as a side-effect of wartime rationing. Centralised bureaucracies and governments are not intrinsically inimical to an ecologically sound society. However oppressive ancient Egyptian society or the Mughal empire or the ancient kingdoms of Polynesia may have been from the point of view of those at the bottom of the social hierarchy, they were ecologically stable, and it was the exercise of authority, often backed by an hierarchical order of priests, which kept them that way. It is the unique combination of hierarchy, centralised power, industrialisation, and the almost universal acceptance of an economic philosophy which regards ethical considerations as beyond its scope, which has landed us where we are in relation to the environment.

Bookchin's distinction between 'environmentalism' and 'ecology' is a useful one; and in drawing attention to the fact that if the ecology movement stops at mere reforms in pollution control and conservation, without dealing radically with social relationships, it will merely act as a safety valve for the existing system of natural and human exploitation, he makes an important contribution to the attainment of a sustainable world. It is in forcing his own particular Utopia upon us, of communities which are not only decentralised participatory democracies, utilising renewable forms of energy, but also libertarian, that he seems to be using ecology for his own purposes. It may be that a society in which interpersonal relationships were transitory and children were a collective responsibility and the old enjoyed no respect from the young nor felt particular responsibilities towards them, would achieve a harmonious relationship with nature, but it remains to be demonstrated. Tribal societies are usually the opposite of this and experiments in this direction have a poor record of survival.

Closely related to Bookchin's claim that the roots of the ecological crisis lie in the domination of some humans by others is the central claim of eco-feminism, that they lie more particularly in the domination of women by men, and that, as culture is to nature, so through most of human history has been the relationship between men and women. Baconian science subjected both. Robert Boyle expressed the metaphor in 1661: 'For some men care only to know nature, others desire to command her, and to bring nature to be serviceable to their particular ends, whether of health, or riches, or sensual delight' (cited in Merchant 1980, p. 89).

The notion that the universe was made for man, and that his destiny is the conquest of nature through scientific rationality is seen as the corollary of the identification of nature as feminine and the idea that women, because of their reproductive functions, are somehow 'closer to nature' than men. 'The basic premise of Eco-feminism', says Ariel Salleh (1988, p. 26) 'is acknowledgement of the parallel in men's thinking between their "right" to exploit nature on one hand, and the use they make of women on the other'. From this it follows that 'the current global crisis is a consequence of the traditional exclusion of women from patriarchal institutions, the most dangerous of these being "science", which replaces religion in our time as ruling myth'.

As women do gain increasing access to power in academic and political structures, in business corporations, and in bureaucracies, this argument will depend on resulting improvement in the global crisis. Meanwhile, the most urgent critics of eco-feminists are often feminists of the more ordinary kind who are not flattered by the notion that female mentality is essentially different, more earthy, and by implication, less rational than that of men. By associating women so strongly with biology and nature, there is danger of playing into the hands of the opposition, and assuming that biology is fate. Most feminists would argue that a distinctive woman's culture is something which has been created by the suppression of the past, like the culture of slave societies, and see it as something which, far from being glorified, needs to be changed as a means of changing an emotionally impoverished male culture, to the ultimate enrichment of the culture of the species (Eckersley 1989).

There is, nevertheless, much fertile common ground which is shared by deep ecology, social ecology and eco-feminism and which is consistent with the Gaia hypothesis. Bookchin emphasises that the distinguishing feature of the human species, its rationality, is no less a product of the evolutionary process than other aspects of evolutionary diversity and excellence such as the flight of birds, the social instincts of ants and bees, the industry of beavers or the navigational know-how of whales. 'This manual we call "Nature" has produced a manual we call "Homo Sapiens", thinking man, and more significantly for the development of society, "thinking woman"' (Bookchin 1987, p. 27).

This 'second nature', implying the unique responsibility of the human species for the future direction of evolution, is as soundly based, he says, in biological principles, as is the process of 'first nature' which proceeded it. If Bookchin is right, and if we accept that the human race is part of Gaia, then it is valid to argue that our interference with nature is natural. This view is made explicit in the theory and prac-tice of bioregionalism, a philosophy which draws heavily on the anarchist tradition of Peter Kropotkin (1842–1921) and looks for inspiration to the cultures of indigenous peoples and the more successful radical communities (not noted for their libertarianism), such as the Amish. The central idea is the notion of the bioregion, a piece of the earth's surface defined in terms of 'biotic shift', or percentage change of species from one place to another. Thus if 15–25 per cent of the species in one place are different from those in the next place, on the other side of a watershed, perhaps, or on the plains as opposed to the hills, then the places are in different bioregions. They probably also have different climates and different soils. Tribal regions occupied by indigenous peoples are held to be bioregions.

Within such units, the common interest of the human inhabitants, ensured by the boundary criteria, will facilitate the degree of consensus required for a functioning partici-patory democracy. Bioregionalism is defined by its pro-ponents as something which 'goes beyond ecology in its enfranchisement of other life forms and its respect for their destinies as intertwined with ours' (Mills 1981). It thus contains the sentiments of totemism, though not its discipline, and its rhetoric combines the romantic with the apocalyptic. It seeks to revive 'bioregions' some of them the pre-nation-state kingdoms of Europe which form centres of modern separatism, and to use them as units of reformed organisation. They will emphasise 'harmony rather than dominance or defence, co-operation with nature and social management rather than central control' (Mills 1981). National states are seen as increasingly on the defensive, concerned to achieve greater degrees of control, and prone to excess. 'The centre will not hold, though. Inflation will make the bureaucratic, police and military glue required to bind things as artificial as nation states together prohibitively expensive' (Mills 1981, p. 4).

Unlike social ecology, however, which shares its common anarchist basis, bioregionalism finds room for a spirituality which arises from a reverence for life to which humanity is inextricably linked. It thus forms a bridge between two of the elements of post environmentalism. The preservation of whales, for example, is on the bioregionalist agenda, not 'merely because they are magnificent creatures, so awesome that when you see one close from an open boat your heart roars; we want to save them for the most selfish of reasons: without them we are diminished' (Dodge 1981, p. 10).

There are currently some 100 bioregional groups in the United States, and doubtless many others in Europe, Canada and Australasia, which represent a partial withdrawal at a personal and community level from the consumer society. Politically they are non-violent and optimistic, confident of the progressive future withdrawal of the nation-state from the areas of life which really matter as the military and industrial force which holds it together become too expensive. Meanwhile the cells of a new polity in the form of local and regional initiatives movements like the 'Land care' groups which spring up in the wake of the rural debt crisis in Australia, can develop and multiply within the body of the dying industrial society. Indigenous peoples, finding themselves denied the affluence which beckoned them into the cities, increasingly reassert their own cultural values and claims to their lost lands. Frustrated majorities are likely to turn to local environmental issues, in which they can experience empowerment, as an antidote to despair.

In this context the present gap between the theories of deep ecologists, social ecologists and bioregionalists, on the one hand, and practical democratic politics, on the other, will probably close, but there are serious problems to be overcome if this is to lead to political success. Deep ecologists are probably wasting their time if they expect majorities to surrender their limited freedom for some 'eco-fascist' future which attempts to save species at the expense of the established social and political values of democratic society. The same applies to proponents of a 'no-growth' philosophy which does not consider, in local detail, the consequences of closing down or cutting back particular economic activities and creating others. Ecological piety on the part of affinity groups and social movements is no substitute for destroying the environment to its position as an essentially political issue over which people will disagree, but for which solutions can be found without bloody revolution. For success in the democratic arena, reformers will need to overcome the stigma of negativity and of privilege which it implies. To argue, as some conservationists do, from the security of a guaranteed personal income, that 'the environment' is more important than jobs, or social welfare, will not win votes. Neither is it wise to take comfort from the manifest inability of present economic policies to deliver the prosperity they promise and to hope for political opportunity in the aftermath of social calamity and political disintegration.

One of the most politically significant aspects of the Gaia hypothesis is that life, as such, is unlikely to be threatened by whatever we do. What is at risk, if we continue to degrade the environment at present rates, is a tolerable society. To ensure the survival, not only of a diverse flora and fauna, but also of a democratic egalitarian and peaceful society, it will be necessary to demonstrate, that environmentally sound methods of industrial production, soft transport technology, and sustainable agriculture make economic as well as ecological sense. Their political sense will then be self-evident.

Suggestions for further reading

At the time of writing Chapter 6 the results of the conference held in Britain (Wadebridge) on the Gaia hypothesis had not arrived in Australia, but they are available from *The Ecologist*.

My colleague Frank McGregor of the University of Adelaide's History Department tells me that the seventeenth-century digger Gerard Winstanley is of more significance in the history of environmental ideas than Jacob Bauthumeley, the proto-deep ecologist. Winstanley's selected writings have been edited by Christopher Hill (1983). Keith Thomas's two works, *Religion and the Decline of Magic* (1971) and *Man and the Natural World* (1983) stand out among the rest for their exhaustive erudition and as a constant reminder that in the world of ideas very little is genuinely original.

The literature of deep ecology, eco-feminism and social ecology grows apace. Some of it is contained in the proceedings of the Ecopolitics IV conference held in September 1989 in Adelaide. Warwick Fox has a new book to be published in 1990, *Toward a Transpersonal Ecology*, which promises to be the clearest and most developed statement yet of the deep ecology position. Other recent discussions are those by Lorna Salzman of the New York Green Party, Glenn Albrecht and Robyn Eckersley in *New Directions: Ecopolitics IV proceedings*, edited by K.F. Dyer and John Young (1990). Ariel Salleh's 1984 article 'Deeper than Deep Ecology: The Eco-feminist Connection' (1984) has been followed by further discussion by Plumwood, Warren and Zimmerman. Jane Yett produced a useful bibliography on 'Women and their Environment' (1984).

Common themes of much of the literature discussed so far are the vision of an ecologically sane society some time in the future, and the urgency of action now, if disaster is to be avoided. What seems to be lacking is guidance about the practical steps needed to achieve a transition from the present state of society to one of sustainability.

If this transition is to be achieved without the destruction of democratic values and freedoms and if it is also to happen within the next generation or so, then it is unlikely that it will be the result of the acceptance, by majorities, of the kind of ideas discussed in the last chapter. In the short run at least, environmental reformers will need to understand the links between their special concerns and those which concern the majority of voters. Majorities will need to be persuaded that, far from being a 'single issue', the crisis of the environment is the whole, of which the social crisis, the economic crisis of industrial society and the 'third world' crisis are manifestations.

So the next chapter starts by arguing that the environmental crisis is not something which transcends existing political parties or ideologies. It shows how environmental reformers are forced by the logic of their own ideas to become concerned about social issues and international problems in conventional terms. It examines some of the ideas of the left and right for their compatibility with a sustainable future. If the society of the future is to be not only ecologically sane, but socially just and democratic then it should be careful not to squander any of its ideological inheritance which may be of value.

Post environmentalist governments which are consistent, and therefore democratic, will need oppositions and alternative governments whose ideas they can tolerate, and which will not get the planet into trouble.

Chapter Seven

Finding common ground

The argument so far leads to the conclusion that the reforms required for the establishment of a sustainable world will have to begin in the wealthy countries which dominate the world economy and either directly or indirectly determine the direction of economic development in the rest of the planet.

In industrial democracies, political solutions to environmental problems are likely to be achieved by an alliance of either 'left' or 'right' with a growing but politically confused environmental movement. Environmentalists, or 'greens' (a name they have mostly come to terms with) would be foolish to claim, as some scientists did in the 1950s, a moral or ideological neutrality for themselves. Remedies for environmental problems are likely to be consistent not only with the personal taste and circumstances of those who promote them but also with their historical analysis of where things went wrong. Hugh Stretton (1969, pp. 52–9) pointed out that there are left-wing and right-wing analyses of wars and revolutions which suggest consistent strategies for avoiding them in the future. In the same way environmental reformers of left-wing and right-wing views are likely to differ over the causes of, and therefore the remedies for, the crisis of the environment.

In addition, they will perceive the crisis itself differently and differ over what it includes and the relative importance of its constituent parts. The crisis we are considering is a network of multiple relationships which continues to develop, but the diagrams on pages 142 and 143 represent typical alternative interpretations and the policies to which those interpretations naturally lead.

'Neither left nor right but in front' is thus a useful slogan but one which is more rhetorical than factual, and ignores problems of which thoughtful reformers are painfully aware. Greens are worried about the middle-class nature of the movement and the contradictions this implies. They cut down no trees themselves, but they read newspapers voraciously, love their personal computers, and know that their careers, if not their personal lifestyles, are inseparable from a culture which is predicated upon such things as the devastation of the Amazon rainforest and the increasing use of electricity, leading to acid rain in Europe or more nuclear power stations in Britain or more dams in Tasmania. When they take a 'radical' stand in defence of the environment they are accused of putting their personal taste for scenery or solitude before the need for others to be employed in whatever dull jobs the industrial system can provide.

'Greens' are aware of the fact that as well-educated people they have never been in danger of involuntary long-term unemployment. Comfortably ensconced in tertiary education, or in the welfare or creative professions, their jobs are usually stimulating and intellectually satisfying. They can usually pay their rents and mortgages without anxiety. Few of them know what it is to lie awake at night worrying about marketing or cash-flow problems.

The result of such contradictions is that within the 'green' constituency in every industrial society there is tension and sometimes conflict between reformists and radicals. As with earlier movements such as liberalism or socialism, reformists believe that goals can be best achieved by working through the existing institutions of society, by existing administrations even, by harnessing existing forces, working with the grain of society rather than against it. Radicals believe that no good fruit can come from a rotten tree and sometimes almost welcome impending disaster as a precondition for a new start.

Ideological differences between reformists and radicals are complicated by fragmentation. Green political movements in every country contain people who retain membership of single-cause organisations. Their activities vary from the deliberate illegality of the Californian 'monkey wrench gang' which 'decommissions' bulldozers and puts spikes in trees to sabotage saw mills, and the 'snowball' campaign in

Causal analysis

Effect

Remedy

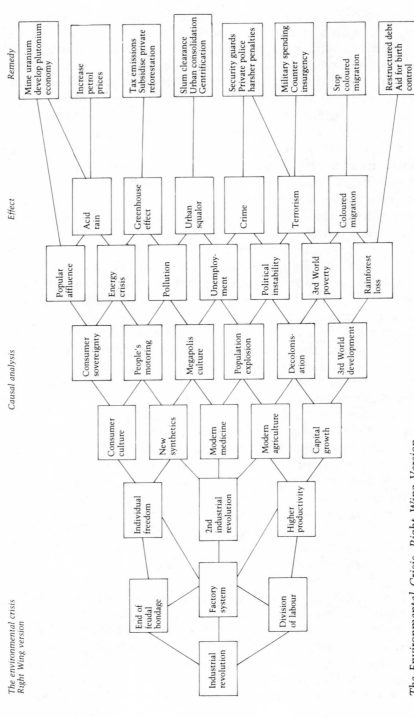

The Environmental Crisis. Right Wing Version

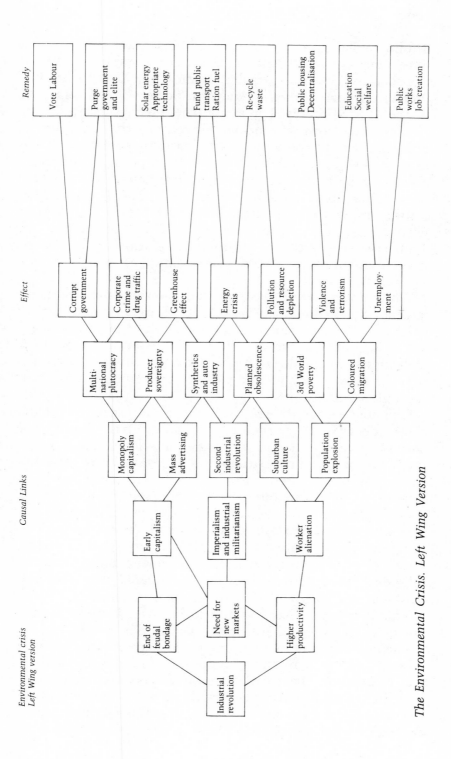

The Environmental Crisis. Left Wing Version

Britain which aims to get its members arrested for cutting wire fences around American defence establishments, to Conservative and 'respectable' bodies like the British Bird Protection Society or the Audubon Society and Sierra Club in the United States. For many, the 'green' slogan; 'Think globally, act locally', means preventing the dumping of nuclear waste in the vicinity of a particular English village, or stopping the woodchip industry at a particular creek in Tasmania, protesting against oil exploration off the northern Californian coast or demonstrating against a new airport runway in Frankfurt. Activity in local campaigns overlaps with a vast network of causes, both soft and good, like Amnesty International, animal liberation, Greenpeace, vegetarianism, Men of the Trees, feminism and the peace movement.

While potentially a movement of enormous strength, the 'greens' are thus characterised by nothing so much as their diversity. The movement recognises that it must think and act in a global context, but it has so far had trouble in developing an ideology of the kind which has provided a basis of consistency for the political movements of the past with global ambitions.

This ideological imprecision may be an aspect of middle-class identity. One sociological study has shown that commitment to the organisations and movements which made up the green constituency is rarely an act of rebellion. Rather, it is an extension of professional commitment to human welfare, frustrated by the constraints of industrial society (Parkin 1968). Action in favour of such causes is a way of asserting moral personality, and the common denominator which unites 'greens', is not, in spite of appearances, class or local interest, but an understanding, however differently personal priorities may be ordered, that politics is less a matter of social mechanics than of morals.

This conviction has the political consequence of leaving greens open to the charge that they do not see the politics of the environment as anything to do with conflict between classes, only as something to do with humanity versus the rest of creation. They are therefore accused by members of existing political parties, especially those of the Left, of not having worked out ways in which proposals for environmental reform such as closing down nuclear power stations or taxing petrol will effect the old and the cold, those who may lose their jobs, or wage-earners who drive long commuting distances in their cheap-to-buy-gas-guzzlers. They are accused of thinking of the environment in a narrow preservationist sense which excludes the workplaces, city centres, and housing estates and consists only of pristine mountain ranges, rainforests, marshland, or the hedgerows of the English countryside. While the inner-city areas of Britain suffered the successive blights of 'slum clearance', unemployment, poverty and social violence in the early 1980s, says one critic, 'the mass of the environmental lobby ... was concentrating its not inconsiderable resources upon protecting hedgerows, butterflies and bunny rabbits' (Weston 1986, p. 12).

Some of those who seek to preserve hedgerows and bunny rabbits may do so for relatively trivial reasons, even while supporting the national economic policies which lead to their destruction. Their reasons do not therefore form part of a coherent environmentalist ethic. There can be, nevertheless, radical implications to even the safest hedgerows and bunny rabbits campaign if it is not merely the action of removing hedgerows which is questioned, but the ideology which justifies the action — the right of either profit-based capitalism or growth-oriented socialism to do what it likes with an important part of nature. To question, that is, the assumptions which make industrial society possible. Others want to preserve hedgerows from a point of view which would disappoint those who think that unemployment and poverty are more important. It arises from a rejection of the anthropocentric view of the universe and the belief that it must be respected and is derived from the ideas of Aldo Leopold and the more recent deep ecology movement.

It may be significant that the movements inspired by deep ecology and the more radical philosophies have developed chiefly in Scandinavia, California, Germany and Australia, where areas of genuine wilderness or forest are to be found, contrasting starkly in the last three cases with some of the most polluting examples of industrial development and urban conglomeration to be seen anywhere, much of it recent. The result is that though the philosophical basis of deep ecology is becoming rapidly more sophisticated than that of old-fashioned nature preservation, its political implications for humans are not dissimilar. Its logic leads to an extreme position either on the Right or the Left. This can be illustrated by comparing the careers of two of the most notable environmental radicals of recent times.

David Brower of California is seen by his supporters as a consistent champion of the environment against the compromises of the political 'realists' who have been successfully corrupted by the need to achieve 'mainstream' credibility. To others he is one who, by preferring the rights of grizzly bears, trees, condors and local eco-systems to those of humanity, has denied humans the spiritual identity which makes them, at least in their own estimation, distinctive among species. Some would argue that it is only the uniqueness of humanity among species which enables people to embrace the kind of new morality which is needed to ensure the establishment of a sustainable society.

For seventeen years Brower was director of the Sierra Club, founded in 1892, and the society which has probably done most on the American continent to make radical preservation respectable. In 1969 he was asked by the Board to resign. It cited 'general intransigence, wilful failure to follow Board directions, unauthorised expenditures', as the reason. His biographer, John McPhee (1971), portrays his defeat as a triumph for the nuclear engineers, real estate agents, publishers and lawyers whose image had been enhanced by membership of the club but who did not share its vision.

Brower then founded Friends of the Earth, a more radical and less parochial but much poorer organisation. He refused to be much troubled by its poverty while the environment was being destroyed at such a rapid rate. 'It's nice to be in the black', he said, 'but more important for the world to be in the Green' (*Mother Jones*, November 1986).

Friends of the Earth has cast a wide net, with branches in other affluent countries, and has conducted inspired campaigns against acid rain, nuclear arms, and James Watt, President Reagan's Secretary of the Interior, whose immediate expectations of the millennium seemed likely to become a self-fulfilling prophecy. Single issues thus dragged the organisation, not without a little screaming, into the arena of national politics, defence and development.

Brower retired as president of Friends of the Earth, but

continued to suffer from what his critics saw as 'Founder's Syndrome', the inability not to interfere. He mobilised his faithful to take a radical line in the 1984 presidential election campaign and support Jesse Jackson or Barry Commoner as a Democratic presidential candidate. This effectively placed such conventional political issues as foreign policy and racial equality on the green agenda. The FOE leadership preferred to throw the weight of the organisation behind Walter Mondale, who, though no particular champion of the environment, had, they thought, a better chance of defeating the man whom they regarded as the greatest threat to the environment, Ronald Reagan.

Political 'realism' and the notion of making not just a point, but a difference, dictated a move of headquarters from San Francisco to Washington in November 1985, a move which manifested the differences between Brower and his opponents and led to his resignation from the Board of Directors. Brower remains active within the Sierra Club, attempting to make it determine its priorities in order of environmental importance, with nuclear power and the arms race now at the top of his list of threats, as opposed to a policy of giving priority to those campaigns, usually of a more local nature, in which victory is more likely.

His own priority is now a new organisation, the Earth Island Institute, based on a crowded office, full of youth and zeal, in a San Francisco attic. Its interests are global and are thus forced to be conventionally as well as environmentally political. Its publication, the *Earth Island Journal*, reports, for example, on Chernobyl, the militarisation of Central America, and attempts by mining companies to extinguish the land rights of indigenous peoples. These are reported as environmental issues, but the writers do not shirk the responsibility of taking sides. Brower editorialises:

Objectivity is a threat because people think it exists and tend to believe those other people who claim to be objective and therefore free of bias . . . if you are going to try to be objective, you must be subjective in selecting what you are going to be objective about. So you start off on the wrong foot, and that wrong foot is attached to everything else in your universe' (*Earth Island Journal*, Fall 1986, p. 2).

His radicalism is something which relates primarily to the dichotomy of nature and man, and only incidentally, though of necessity increasingly, to the dichotomies between some members of society and others. Earth First! radicals maintain that social injustice is a side-effect of a more general immorality, 'a world based upon exploitation, not only of resources, but of humans, and domination, not only of Nature, but of classes' (*Mother Jones*, November 1986, p. 33). Socialists with environmental sympathies would question this. They would say that while the end of class exploitation would allow the possibility of an end to the exploitation of nature, the reduction of demands on nature would not necessarily mean a reduction of class exploitation. It could well increase it instead. Conditions for the poor unaccompanied by a redistribution of wealth in a society which took less from the environment, would get worse rather than better. Increasing radicalism in the man versus nature debate can thus be an alternative to a commitment to egalitarianism.

Brower's position is thus typical of those whose logic leads them in this direction even though their actions and the causes they support land them in the company of liberals, democrats and socialists. 'Friends of the Earth made the Sierra Club appear reasonable', he says, and 'Earth First! makes F.O.E. appear reasonable. Now we want some people to make Earth First! appear reasonable' (*Mother Jones*, November 1986, p. 33).

David Brower can be understood as one whose environmental radicalism has led him towards policies which reduce the number of enemies he has on the Left and increases the number on the Right. Such are the consequences, for example, of opposing nuclear power or the sending of arms to the Nicaraguan *Contras*. His European counterpart, in many ways, is Rudolf Bahro, who began political life as an orthodox citizen of East Germany, became its constructive critic, fled from it, and became one of the best-known critics of centralised state power, especially socialist state power, in Europe. He is also a leading exponent of the European green movement, broader in scope than its American counterpart, and because its practitioners have been involved in the day-to-day business of political debate and legislation it has been forced in the direction of consistency.

Bahro arrived in West Germany in October 1979, after his release from East Germany, where he had been imprisoned since August 1977, ostensibly for espionage, but evidently because of the publication of *The Alternative in Eastern Europe* (Fernbach 1979), a scholarly critique of East German bureaucracy and government, and also because of press and television interviews which he had given to the West German media. He immediately became involved with the green movement and was a founder member of the political party which achieved its most spectacular success in March 1983, when it won twenty-seven seats in the Bundestag.

Debates within the party centred on the desirability of entering at all into the national parliamentary arena, and, once in that arena, on the alternatives of standing alone or of forming a coalition with one of the existing parties. If a coalition were to be formed, which party should it be formed with — the conservative Christian Democrats (CDU), or the left-liberal Social Democrats (SDP)?

Bahro rapidly gravitated, not to the *left* of either, but to a position based on the paramouncy of the environment over the immediate interests of any particular section of the human race. His developing stance placed him in ideologically close company with the David Brower of twenty years previously. The target of his most scathing criticism has been what he calls the *realos*, people who in his view have a strategy of actively assisting with repairs to a system which is best left to collapse. The achievement of reforms such as the fitting of anti-pollution devices to automobile exhausts is seen by Bahro not as the tangible rewards of 'realistic' policies, but as palliatives which must be attacked because they have the effect of postponing a catastrophe which is both inevitable and desirable.

Such a philosophy is uncomfortable to a party tasting significant electoral success, and Bahro, like Brower, became, depending on your point of view, either the representative of developing extremism, or a rock of consistency standing against a current of compromise which co-opted the green movement and made it impotent as a force for change of any significance. He eventually resigned from the West German Green Party, ostensibly because of a party resolution which condoned animal experiments (though only in cases where it could be shown that they had the

effect of saving human life). This, he said, 'revealed the principle by which human beings are exterminating plants,
animals and finally themselves'. Not that political realism
was absent, but it was a realism directed towards a new kind
of constituency which had not been seen before as relevant
to party politics. 'After what happened yesterday, am I still
supposed to try to win serious animal protectionists over to
our side?' he asked (Bahro 1986, p. 210).

The continuum from 'realist' to radical within the green
movement is thus one which not only cuts across the left–
right political spectrum, but extends the scope of politics
itself to include intergenerational conflict and the rights of
non-human species. The movement therefore includes those
who despair of the politics of parliamentary democracy as a
means of initiating change. 'Greenpeace', for example, uses
the strategy of selecting issues likely to receive maximum
public attention in order to make the largest possible
number of people think about them seriously. Preferring
non-violent direct action to debate, members accept the
duty, Quaker-like, to 'bear witness' by the kind of well-
documented and publicised displays of personal courage for
which they have become famous. They have placed themselves in small inflatable boats between harpoons and
whales, or beneath drums of nuclear waste as they are
dumped into the sea.

Most extreme on the humanity versus nature scale are,
perhaps, the California-based deep ecologists Bill Devall and
George Sessions. As if to chide Schumacher for writing
about 'Economics as if People Mattered', the sub-title of
their book, *Deep Ecology: Living as if Nature Mattered*
(1985), implies that in the last analysis, people don't.

These authors define deep ecology as something which
requires fundamental social change, but they argue that it
does not require a political programme. It is not, they say,
an attempt to add another ideology to an already crowded
field, nor does it call for the overthrow of existing governments. Change will be wrought on an individual basis by
those who continue to search for new ways to 'liberate the
ecological consciousness'. Warwick Fox (1989), the leading
Australian exponent of deep ecology takes the argument
further, and, implicitly at least, into the realm of politics.
He urges a redefinition of this branch of eco-philosophy as

transpersonal ecology. This means that in place of a personal identification of self with family, friends, ethnic grouping, species and so on, transpersonal ecology means:

that we strive . . . not to identify ourselves exclusively with our leaf (our personal, biographic self), our twig (our family), the leaves (our friends), our minor sub-branch (our community), our major sub-branch (our race), our branch (our species), and so on, but rather to identify ourselves with the tree. This necessarily leads, at the limit, to an impartial (but deeply felt) identification with *all* particulars (all leaves on the tree). (Fox, 1989).

Actions which follow from this philosophical position, he argues, include not only 'treading lightly upon the earth' but 'also actions that respectfully but resolutely attempt to alter the views and behaviour of those who persist in the delusion that self-realisation lies in the direction of dominating the earth and the myriad entities with which we co-exist' (ibid).

In Europe, Bahro points out that, historically, the foundations on which new cultures have been based, following the decay or collapse of the old ones, have belonged 'to those strata of consciousness which are traditionally described as religious'. He expects that, as with Christianity in the context of the floundering Roman Empire, the international proletariat will create a higher religion, stemming from a contemporary 'peripheral culture'. He notes in this context the extent of 'development aid' in the form of spirituality currently being given by the Indian sub-continent to backpackers from the spiritually underdeveloped areas of Europe and the USA. Noting also the upsurge of Christian fundamentalism in the affluent countries, he concludes that, 'the contest with the Apocalypse can only be won if this becomes a great era of belief, a Pentecost with the living spirit poured out as equally as possible over all' (Bahro 1986, p. 188).

Even the relatively staid Green Party of Britain is eclectic in its enthusiasm, calling for a recognition of 'the right of all people to enjoy our common heritage, and to accord the right of freedom of religious expression to pagans and to all those who regard Stonehenge as a temple' (Green Party 1986). This resolution of the party conference dramatically underlines the breadth of the green constituency and the difficulty of achieving consensus, for it was the conservative and conventional National Trust and English Heritage organisations which were once again to line up behind the Thatcher government and an army of policemen to keep the 'hippy' convoy of Midsummer Day 1986 moving past Stonehenge as usual. Since then, Mrs Thatcher has proclaimed herself a convert to green values, but this has not so far resulted in a reduction of violence at this annual occasion.

While many 'greens' would question the ecological credentials of an autonomous Ulster or Alsace, and religious scholars may have difficulty in coming to serious terms with deep ecologists or British pagans, these movements are evidence of a felt need for a spiritual and cultural link with particular parts of the earth rather than the planet as a whole to provide an overriding reason for good stewardship.

What seems to be missing is an understanding of the mechanics and disciplinary implications of such links in societies which actually live by them. In spite of the young backpackers in Nepal, there is something Eurocentric if not provincial about this search for a new spirituality. More consistent, and more to the point for the purpose of maintaining a sustainable society is the ethical system, and its religious framework, of some of those peoples whose religious belief has been part of that conservatism which resists Western consumerism and remains as a flickering flame which never quite went out during the decades or even centuries of colonialism. It is likely to be revived by the discovery of new ecological reasons for having confidence in those aspects of traditional culture which are not yet discarded, and could make a useful contribution to a global ethical consensus. To use Warwick Fox's analogy, it is specifically because of the definition of self in terms of adjacent leaves on the tree, rather than the broader, but inevitably less powerful identification with the tree as a whole, that results in an ecologically responsible ethic.

Central to the ethical system of all peoples whose relationship with the land has remained close is the attachment of great importance to kinship and the obligations which go with it. These involve primary obligations to particular ancestors, not merely to ancestors in general, the duty to honour them by using the land wisely for the continuous fulfilment of obligations to particular members of contemporary society. This of itself is a guarantee that the method of use will be sustainable so that resources can be passed on, undiminished in value, to children in the future. It is such finite ground rules which give rise to principles of loyalty and reciprocation, of obligation to ancestors and to children in general, and to less tangible feelings of identity with particular landscapes and regions and with the specific spirituality which defines cultural identity. Ideology thus arises from ground rules of environmental management, not, usually, the other way round, through exhortation or inspirational 'conversion'.

Transposed into the context of modern industrial civilisation, this makes the extended family an excellent basic unit of a sustainable society, and though less reliable, the nuclear family is probably better than none at all. Greens should therefore be wary of joining in the current enthusiasm for its demise (see, for example, Davis *et al.* 1980). Those who preach the virtues of the small co-operative commune or who call, as Bahro does, for a new 'Benedictine order' of both men and women who will 'give space to the uninhibited development of sensuality and sexuality' (Bahro 1986, pp. 90–1) as the social basis of the new order would do well to study the well-ordered life of a communal society based on kinship, such as the Fijian *mataqali* or land-owning unit. It has a complex code of personal conduct, of familiarity and avoidance, deference and responsibility. The rules have developed as a balance between the disruptive and cohesive forces in society and between society and its local environment.

Christianity has often succeeded best in such societies when it has been successful in arresting the process of cultural disintegration at the hands of the forces of colonialism. Often it has rescued traditional morality and customary behaviour from the disruptive consequences of transition from a 'closed' to an 'open' society, in which change becomes an acceptable consequence of human activity (Jarvie 1964). Nacalieni Rika (1974), for example, a Fijian teacher speaking at a conference on development, argued that the obligations of kinship were too valuable to be thrown aside as obstacles to progress. Urbanisation and the institutional isolation of the two-child family had led to the deprivation of wider associations and threatened to undermine the morality which was needed to sustain them.

The Californian Earth First! deep ecologist who seeks a

spiritual basis for a desire to protect the environment thus stands, in some respects, on the opposite side of a moral divide from millions of people in poor countries who have the spiritual confidence he or she seeks. These millions would certainly share many Earth First! criticisms of industrial society, but they would probably reverse the Earth First! order of priorities. They would place particular people first, then categories of people defined in terms of relationships, and including living, dead and future generations, then other species, hence the eco-systems of which people form a part, and without whose health human obligations cannot be fulfilled.

Npare Hopa (1990) a Maori scholar from New Zealand, notes the correspondence between Maori cosmology and some of the most recent and radical environmental philosophies. The individual Maori, as in all Polynesian societies, is embedded in society and society is embedded in nature and dependent on cosmic forces. Humanity and nature therefore do not stand in opposition and do not have to be understood in different ways. Natural phenomena are apprehended in terms of human experience and human experience is understood in terms of cosmic events.

The entire cosmos is thus kinship writ large, a cosmic family that is mirrored in the social environment. Speaking of the Gaia hypothesis, she says: 'They might just as well have invoked the creation myths of many tribal people including the Maori myths of Papatuanuku, the Earth Mother'. Lovelock, in her view, has provided scientific support for something which 'the astronauts and countless men and women on earth before them realised intuitively that individual life and identity is part of a living earth.' 'Deep Ecology', she says, 'is thus not new.'

Since the 'protesting 70s' Maori people have begun to think globally and to act locally in their bid for self-determination, social justice, ecological balance, self-sufficiency and spirituality. They are using the framework of the Waitangi Tribunal, set up in 1975 to examine Maori claims against dispossession in the light of a reassessment of the legal status of the 1840 Treaty of Waitangi by which New Zealand was ceded by a number of chiefs to the British Crown, to demand a share in the wealth and management of the natural resources specifically mentioned in the treaty as remaining under Maori authority. These include certain fishing grounds, including offshore fishing grounds, areas now mined for coal, and forests, which they wish to develop according to their own labour-intensive, small-scale economies, in the context of their reviving tribal structures and values.

Hopa asserts that the claims which Maoris have placed before the Waitangi Tribunal are not just intended to adjust the current social and economic balance and the problem of Maori dependency, but to re-establish and revive their cultural understanding. It was never truly lost, in spite of loss of lands, loss of population, social division, war, urbanisation and the marginalisation of the past. Now it has become integral to the renaissance of Maori society.

Incorporating the contribution of such cultures into the ethics of a sustainable society would mean interpreting the slogan 'Think globally, act locally' in a non-prescriptive sense. Thinking locally as well, and acting personally, can provide a reliable foundation for a larger and more culturally inclusive morality.

Conservatism as it exists in affluent Western societies is also a source of some ideas which could contribute to an ethically consistent attitude to the environment. The monetarism which dominates the morality of the northern hemisphere at present may be the antithesis of the spiritual values of Fiji, or of the Burma which so impressed Schumacher in 1955, but there are points at which the values of kin-based societies correspond with those of conservatives in countries like Britain and the United States.

Edmund Burke (1790), the source of many of the more attractive aspects of modern conservatism, described the basis of government and society in moral terms which echo the principles of many kin-based pre-industrial societies. This basis was

a partnership, not merely between those who are living, but between those who are living, those who are dead, and those who are to be borne. Each contract of each particular state is but a clause in the great primeval contract of eternal society, linking the lower with the higher natures, connecting the visible and invisible world, according to a fixed compact sanctioned by the inviolable oath which holds all physical and moral natures in their appointed place.

Conservatives who inherit this view have a contribution to make to green morality which is distinct from the reverence for wilderness with which they are generally credited. They often believe, like their 'Third World' counterparts, that possession of land carries with it a responsibility for its proper management in the interest of both present and future generations. They believe in the importance of locality and community, and, like bioregionalists, oppose the expansion of the centralised power of the state, more especially if the state is in socialist hands. They dislike waste, and like to think of national finance in pre-Keynesian terms analogous to good housekeeping and paying bills on time. Consumerism of the popular kind is distasteful, indeed conspicuous consumption by those unused to it is what conservatives call 'vulgarity'.

In practice, however, modern conservatives are confused about the environment. Their elitism has made them fatally vulnerable to seduction by the ethos of industrialism, if only as a means to the retention of power, which is the primary conservative responsibility. The first to profit from the opportunities of industrialism were those members of the British aristocracy who owned coalmines, displaced tenant farmers to house factory workers or patronised inventors. In Germany, the United States and Japan, conservatives have been prepared to risk an alliance with industrialised militarism to ensure political survival.

For the reasons explained in Chapter 1, a commitment to inequality in the context of industrial society makes it necessary for conservatives to favour a policy of economic growth and consumerism in order to contain the social discontent which inequality would otherwise create. As a recent British pamphlet puts it 'The danger is that the Tory party may be seen as having been taken hostage by the "Ah well, that's progress" lobby, consisting of industrial polluters, farming vandals, Green-belt grasping house-builders and city centre property developers'. The Conservative Party does not receive credit from the voters for 'having a green thread running through our thinking, nor do we deserve to, for sadly, it does not' (Patterson 1984).

There is thus a contradiction, between the ethics so well expressed by Burke, together with the love of countryside, of unspoilt mountains and forests, in the hearts of many conservatives, and the economic, foreign and defence policies favoured by contemporary conservative governments.

Conservative environmental concern therefore has to remain for the most part parochial. Conservatives campaign in favour of favourite causes, rare species, special bits of countryside, hedgerows or old buildings. Concerned British Conservatives are reminded, that 'there are 1 million hunters and hunt followers and 3 million anglers, all with a special interest in preserving the countryside'. The 'conservation card' is recommended as an invaluable means of thwarting the attempt of the Liberal Democrat and Social Democrat parties to drain off support from the expanse of soggy "wetlands" at the centre of British politics' (Patterson 1984). Such cynicism betrays either a shallow aim of exploiting the environment as a party-political issue or a deep misunderstanding of the scope and interrelationships of the environmental crisis. Further inconsistencies arise when green-minded Conservatives think about such matters as foreign policy. However much they mind the removal of their favourite hedgerows, English Tories are likely to baulk at withdrawal from the European Economic Community, to whose agricultural policies the destruction of hedgerows can be largely attributed, almost as much as they would at withdrawal from NATO. It is therefore more likely that the 'green thread' embedded in conservative thought will be a means of detaching greens from conservative parties than of detaching conservatives from green parties.

The theoretical difficulties encountered by Labour and socialist parties in broadening their philosophies to address the problems of the environment are different from those which face Conservatives but, like them, they arise from the structure of industrial society. 'Red' greens thus believe that ecological reform is a two-stage process. The first involves equalisation of society. This will make ecological reform possible because only a more equal society will be able to share the necessary sacrifices in a way which will ensure that they are tolerated.

But Labour parties also have their problems when it comes to environmental reform, stemming from their entrapment within the framework of industrialism. Historically, they have sought to combat the social outcome of the alliance between technology and capitalism; but they have fought on ground of the enemy's choosing and it has been a defensive rather than an offensive struggle. Workers forced to accept long hours, low wages, dangerous conditions and boring jobs have gone on strike to improve conditions, get higher wages and, more recently, to reduce hours of work, but since the days of Ned Ludd they have made no protest against the division of labour and the alienation which results from it. The political achievement of the Labour movement as a whole has been to compensate for rather than to remedy the situation in which the part of life worth living is divorced entirely from work.

Because it has been materialistic, the left in countries like Britain, West Germany and Australia has emphasised the importance of national solidarity and central control in order to achieve economic growth more efficiently and more equitably. Often it has succeeded brilliantly, and, when things go well, left-wing governments and parties attract the support of the really big industrial concerns, which need the mass markets which well-compensated trade unionists provide.

To this extent, the development of socialist environmental strategy designed to operate in the context of a centralised industrial democracy has been a blind alley leading to a dead end. To take full advantage of the political opportunities of the ecological crisis, socialism may have to return to the crossroads which it had reached in the last quarter of the nineteenth century.

Charlotte Wilson, author of a Fabian tract in 1886, believed that in Britain there were two potential socialisms, a collectivist socialism which would lead, naturally, to the establishment of a strong centralist administration, and a counterbalancing anarchist socialism, defending individual initiative against it (Wilson 1886). Environmentalists of the left are now seeking to re-emphasise this minority tradition because of its ideological compatibility with such goals as decentralisation, bioregionalism and small-scale economic activity. They point out that the kind of socialism which depends on a strong central administration has been in the ascendant for almost a century, in both Britain and the allegedly socialist countries of Eastern Europe. One result of the popular identification of socialism with state power has been the repeated election to office of governments like those of Mrs Thatcher and Ronald Reagan.

Joe Weston argues that many green ideas originate in now discarded elements of socialist ideology, including the belief in 'natural' limits to human achievement, the denial of class divisions and a romantic view of nature. Preference for small-scale communites, participatory democracy, belief in a 'natural relationship' between people and the earth and even a 'new spiritualism' are elements of radical ideology which go back to the peasant revolts of medieval times, to the Diggers and Levellers of the 1640s, and which are echoed by Robert Owen and the Ricardian socialists of the 1820s and 1830s, by Prudhon and the American populists such as Henry George in the 1880s. 'Far from being a "New Paradigm", [which] transcends the old political framework of left versus right', green politics is part of a long tradition in political thought, older in fact than both Liberalism and Marxism, yet encompassing elements of both' (Weston 1986 p. 25).

In practice, Labour and social democratic governments in recent times have offered alternative management of the capitalist economy, and they have won most popular support when they have seemed better at it than their unashamedly capitalist rivals. This lends force to Bahro's (1984, p. 169) contention that 'the whole right–left question has not the slightest to do with the ecological crisis, because the S.D.P. with the unions behind it is at least the equal of the C.D.U. as the party of expansion' (1986).

Socialists reply by accusing greens of inhumanity. The leakage of radioactive waste from Sellafield disturbs them (socialists say) ostensibly because it caused leukemia in children, but really because it is a threat to the ecological system in the vicinity. How else to explain the lack of concern at the number of people killed each year on the roads? What do socialists want, then? More motorways, policemen and radar traps? Don't they understand that road deaths are a mere side-effect of emphasis on road transport and economic growth? Don't greens realise, (say the socialists) that the environmental problems which really matter, like poor housing and street violence, are aspects of the underlying problem of poverty.

Both sides thus accuse the other of being naive, of the need to make linkages and to see greater complexities understood only by themselves, but greens do share with some socialists at least a similar vision of an ideal and sustainable society. Differences are often those of style and priority. Greens tend to attack industrialism rather than capitalism. Socialists believe that it is not industrialism that is the problem, but who controls it, and that under proletarian local control it makes life better for the majority. When it comes to working together there are sometimes cultural differences, too. David Pepper (1986, pp. 125, 138) complains of the social implications of green radicalism: 'People who are not interested in modifying their personal behaviour and thought patterns to accord with the view that they are merely a sort of animal have been excluded'. He sees this as a major obstacle to an alliance between socialists and greens because 'It is not merely difficult; it is nigh on impossible to imagine Cowley car workers, Sunderland soccer fans, Brixton bomb-throwers or even Cheam commuters fitting into the sickly, sanctimonious tree-hugging communities of Callenbach's *Ectopia* [1978]'.

Pepper's context is peculiarly English, but the problem is a product of the fact that in modern industrial society, class, for that is the difference he is talking about, is determined by how people function as consumers rather than as producers. As consumers, Cowley car workers and Brixton bomb-throwers alike are victims of the shoddiness and rapid obsolescence of mass-produced products and of an advertising industry which creates wants which can never be satisfied. Hope can be sustained only by participation in the industrial workforce at rates of pay which big unions have fought for and won. Historically, factory workers were the enemies of the craftsmen who resisted the formation of mass unions because they were also small businessmen. Craftsmen, together with members of the caring and educating professions now emerge ironically as members of the constituency with which the left must come to terms if it is to be effectively green.

Unionists have traditionally been narrow in their demands relating to the environment, which they have defined primarily as the workplace, but there have been exceptions. Jack Mundey was leader of the 'green bans' movement in Australia, aimed at the attempts of developers to destroy what remained of Sydney's heritage buildings, then at coalmining in the vicinity of beaches frequented by workers and their families. He confesses to some amazement that the unions will make social demands but do not question the overall results of their labour as a contribution to the industrial system as a whole. They remain content to react to economic circumstances at particular places in particular jobs, one at a time (Mundey 1990).

There is thus a task of emphasising broader linkages, of environmental education within the trade union movement and the Left generally, if green politics is to benefit from the practice as well as the theory of socialism. Sometimes this happens as the result of exemplary action. The British miners' strike of 1985 failed because it was successfully depicted as a rearguard action to keep uneconomic pits open. It proved useful to the Thatcher government to have a 20 per cent contribution to the electric grid from nuclear power. However, though defeated, the strike proved a useful educational experience for the left, uniting miners, environmentalists, feminists and pacifists. Gradually it becomes evident in a similar way that the specific environmental problems which effect workers, like lead poisoning from petrol fumes, pesticides which poison the food chain, and so on, are not really single issues at all. Lead in petrol (and road accidents for that matter) raises the issue of public as against private transport. Pesticides raise the issue of agricultural policy and regional specialisation, leading to overproduction of unwanted commodities like grain and butter by rich countries and to the destabilisation of peasant economies in poor ones.

There can be little argument with the contention that for most of the world's poor, in both rich and poor countries, the most pressing environmental problem is poverty. Differences between socialists and greens centre on what to do about it. The method by which it could be most easily conquered in the age in which energy seemed cheap and limitless was to apply, to whatever degree the electorate would tolerate, the principles of state socialism, including wage indexation, compulsory unionism, an elaborate system of social welfare and national health schemes, and graduated income and other taxes. This tended to redistribute the proceeds of growth, but the eradication of inequality and poverty by this means does little in the long run to help the planet. The more effectually equality is pursued within the context of growth, the more the rich countries will consume. More 'primitive affluence' in subsistence economies and more rainforest will be destroyed to facilitate the production of beef to supply the fast food chains of the 'well-fed' state.

In countries like Australia, the United States and Britain the increasing power of the rich, their control of the media and their ability to exploit the laws relating to taxation to their advantage and to move their wealth around internationally has made it difficult to control extremes of inequality. The moderately rich have also profited from tax laws designed to encourage economic growth and the tendency for even Labour governments to emphasise indirect methods of taxation which are hard on the poor, but more difficult than direct taxation to evade.

State socialism, however diluted its form, thus offers a poor framework for the green movement. If socialism is to be of use it will need, therefore, to borrow ideas from its own anarchist tradition, and as environmentalists become conscious of the contradiction between their class affiliations and their hopes for the planet, so left-wing parties and trade unionists will need to understand the advantages of decentralisation.

In this context, unemployment, the social outcome of the alliance between technology and capitalism, becomes the soft underbelly of the industrial system and presents the opportunity for positive socialist–green collaboration.

While energy was cheap, in the expansion decades which followed the Second World War, unemployment rates of 5 or even 3 per cent were enough to topple governments. Now, the best that any 'developed' economy can achieve is around 7 per cent, while figures of between 10 and 15 per cent in Britain, Europe and the United States are seen as something we will learn to live with. These conditions may, as in the past, prove fertile ground for all manner of nastiness, but they may provide an opportunity which can be seized upon as a starting point for a peaceful transition to a sustainable society.

On an individual level, unemployment is an opportunity for the re-education of consumers in the direction of autonomy. Nationally it is the opportunity to create the structure of a less wasteful society. Internationally, the restructuring of wealthy economies will allow for choices between different paths of development by poor countries, freed from the pressure to reduplicate the industrial experience with its bad as well as its good consequences, as a result of the demands of the rich ones.

Greens should understand the relevance of this crisis of industrial society to their concerns, and be prepared, instead of merely prophesying universal doom or calling attention to the destruction of wilderness, to accept responsibility for promoting constructive proposals as alternatives to the economic system which underlies the specific environmental problems which happen to cause them most concern.

As national governments become increasingly impotent, because of their macro-economic and 'defence' policies, to deal with environmental problems except in cosmetic fashion, the initiative naturally falls to local government, and this accords conveniently with the developing anarchist, communal and regional thrust of green politics.

Local governments have proved their ability as the means of initiating change and staging exemplary action. Examples include the world-wide nuclear-free zone movement, the far-reaching initiatives of job-creation, appropriate technology, urban farms and public transport which were introduced before its demise by the Greater London Council. Local governments all over the world have been responsible for strategies such as the profitable recycling of material (in Santa Barbara, California, and Canberra, Australia) local economic self-sufficiency (in Aberfeldy, Scotland) for obstructing the introduction of commercial television (in Suva, Fiji). Most importantly, they have provided the structures and opportunities for the job-creation schemes of many kinds which have been the characteristic response of national governments to the problem of unemployment.

These provide an opportunity for the diversion of resources from the servicing of industrial society towards the development of alternatives to it. Money now spent reluctantly by conservative governments and profligately by socialist governments on providing jobs within existing industries should be spent, especially by local governments, on developing those enabling skills which make it possible for people to employ themselves, pool their skills with those of others to fill real needs and so avoid the need to depend principally on big business or big government for employment.

The difficulty with schemes which emphasise 'training' rather than the receipt of wages as prescribed by the relevant award is that they are seen by the left as a new way of performing the old trick of lowering wages. The right reinforces this view by arguing that lower wages, especially for young people undergoing 'training' are the only way in which more employment can be generated, especially by 'small business' which cannot afford to pay high wages. Both arguments are about half true (small business cannot afford high wages, but big business can, and does very well out of low wages) but both deny that under present conditions there will be fewer jobs in any case, and 'training' which is merely teaching people how to fit into the industrial process is not going to create new jobs as fast as technology eliminates old ones.

Greens should therefore support the creation of jobs in which the emphasis is not on gaining short-term experience of work, earning a wage so as to be able to rejoin the ranks of the consumers, but on a longer-term educational and practical experience which can enable people, if they choose, to achieve a degree of independence from consumer society. Job schemes should be funded to the degree to which they economise in the use of non-renewable resources and energy, and their commitment to resources and energy of a renewable kind. They should be used to develop appropriate technology, to preserve or create skills which enable participants to save money by making or building things for themselves, to create value by renovating old houses or building new ones to energy-efficient designs using recycled or renewable materials.

Such enterprises have the advantage of compatibility with the anarchist, Owenite and even Christian varieties of socialism and also with such ideals as the decentralisation of authority, local participatory democracy, even bioregionalism. At the same time they are active rather than contemplative. For this reason, and because they provide scope for the development of individuality, such enterprises have the valuable capacity to win over the support of the right as well as the left, especially in wealthy countries. Governments concerned to take the current when it serves should provide guidance, encouragement and, when needed, technical advice and finance. The post environmentalist movement should recognise the mix of soft energy, appropriate technology, sustainable economics, communal autonomy and cultural identity as the essential ingredients of a consistent ideology which can harness the diverse energies of a crisis-conscious generation.

In the context of an unprecedented crescendo of public concern and the sudden anxiety on the part of major political parties everywhere to 'capture the green vote', there is danger of being co-opted without being understood, and of a coherent green ideology being dismembered from left and right. If that can be avoided, and if green politics can develop an ideology which demonstrates the relationship between the deterioration of society and that of the natural environment, then its task will be the relatively easy one of persuading people democratically to accept the inevitable.

Suggestions for further reading

This chapter had its origin in California in 1986 and was based initially on the periodical literature of the various green organisations of that state, chief among them being *Mother Jones*, the *Earth Island Journal* and *Co-evolution Quarterly*. Apart from newspaper sources, there are some thoughtful discussions of green politics in Europe, among them 'Stained Greens: Interview with Petra Kelly', in *Marxism Today*, no. 29 (June 1985), P. Wilhelm Burklin's article, 'The German Greens. The Post-industrial Non-established Left and the Party System' (1985), and John Wiseman's 'Red or Green? The German Ecological Movement' (1984).

Green politics in Australasia is discussed in Drew Hutton's collection *Green Politics in Australia* (1987) and in a collection of essays edited by Peter Hay, Robyn Eckersley and Geoff Holloway, *Environmental Politics in Australia and New Zealand* (1989).

Recent developments include the launching of a new left party which seeks to promote green issues while remaining determinedly socialist, and a 'green faction' within the Australian Labour Party.

The Australian Democrats, originally a group of small-'l' Liberals, who broke away from the conservative but so-called Liberal Party, are now attempting, with some success, to identify themselves as Australia's green party.

Pacific island societies, their environments and the problems they face are treated in detail by Randy Thaman and Asesela Ravuvu, 'Ecopolitics and the Destruction of the Pacific Island Environment: A Crisis of Cultural Survival' in Dyer and Young (eds) (1990).

Different political circumstances, constitutional arrangements and degrees of affluence call for different strategies. In parliamentary democracies it is not sensible to imagine a future in which the achievement of power by a 'green' government is the dawn of a new green millenium. Other parties will win back power quite quickly, as they have in the past, but that should not mean the resumption of environmentally damaging policies. Just as some of the best exponents of the principles of state socialism have been ostensibly conservative governments, and modern labour parties have become the ablest champions of big business, so it should be possible to ensure that all parties have sensible long-term environmental policies which will inevitably allow for different ideological commitments, hence different ways of doing things. Environmental protection is now an accepted responsibility of all parties; in future, no government of a 'developed' country is likely to be without a policy of environmental reform with which its other purposes should be consistent. There will therefore be implications for other permanent responsibilities, treasury policy, education, defence and so on. Different kinds of reform can be grafted on to the policies of different kinds of political party, and party statements and papers suggest what is likely to work and what is not, for reasons of political consistency outlined in Chapter 7.

Different constitutional arrangements call for different strategies, but almost none of them will be wasted. Proportional representation and preferential voting are favourable to new parties and make it less necessary to compromise with other parties but more necessary to compromise with other issues like unemployment or defence for which green parties will become responsible.

First-past-the-post systems and presidential systems generally make it essential to influence existing parties, departments and bureaucracies so that it matters as little as possible who is in power. Radical reform is easiest to initiate at local and community levels, creating a new framework within which national government must eventually operate.

The scenario of indefinite onward and upward growth which caused the sense of crisis in the 1960s now seems less and less likely anyway. It seems more likely that the multiple contemporary crises of economic instability and unemployment, of famine and desertification, of climatic change, drug addiction, social violence and disease marks the beginning of the end of the civilisation which will be remembered as industrial. The emergence of green politics as a major force makes the transition to post-industrial society an opportunity for a peaceful and invigorating transformation.

Chapter Eight

The politics of a sustainable society

The decline of the kind of industrial society born at the beginning of the nineteenth century in Britain, then in Europe and America, now in Asia, Africa and the Pacific, has evidently begun already. The 'stages of growth' will be drastically foreshortened in much of the world and in others they may not happen at all. Unless nuclear war occurs, a possibility which has not diminished with the proliferation of nuclear weapons, the question is not whether the structures of late industrial society will survive, such as the corporation, the trade union, the bureaucracy and the nation-state, but how they will change, whether or not the result will be a society which will be sustainable, and whether sustainability, liberty and democracy will prove to be compatible.

Many writers have seen a need for such a transition to be sudden and catastrophic. The hope is that when the crisis comes, through the exhaustion of fossil fuels, Wall Street 'crash' or holocaust, there will be enough converts to the side of ecological sanity to take advantage of the situation. This apocalyptic tradition has much in common with the earlier Domesday tradition of the 1960s, based as it is on the fantasy of standing alone in the smoking ruins, having been right all along.

If such fantasies become a substitute for rational action they can be dangerous. Historically, revolutions have tended to lead to results which cannot be foreseen at the outset. A sustainable future society will be one which builds on those elements of existing societies which have survived what can now be seen as the ecological aberrations of the late industrial period of human history. While the Apocalypse remains, as always entirely possible, it is sensible to plan for a situation in which it does not happen; to assume the possibility of gradual change, beginning most easily at the level of individuals, families and communities, where it is already well on the way in the most highly industrialised societies, and moving from them to unions, corporations, academies, districts, political parties and nations. Failure of Armageddon to eventuate will not then be disappointing and time will not have been lost in anticipation.

Some economists who do not believe in growth regardless and are sympathetic to the problems of the environment nevertheless retain faith in market forces to solve the environmental crisis. As resources like oil and coal, timber, fresh water and wilderness become more scarce, they will become more expensive, and this will have environmentally beneficial effects. Industry will have the necessary incentive to invest in renewable sources of energy and to recycle materials, which will lead in turn to cleaner air, more rational conservationist forms of transport, even greater local self-sufficiency. Most importantly it will provide an incentive for timber, oil, even uranium to be husbanded for the future — the short-term future, anyway. Ownership of resources, whether by indigenous peoples or mining companies, will become more valuable as a commodity than the profits to be derived from immediate exploitation. According to this argument, the belief of Amory Lovins, advanced in *Soft Energy Paths* (1979) that a society which gets its energy right will get its relationship with the environment right will be tested as a direct consequence of scarcity without political intervention.

The trouble with this theory is that it places faith on the ability of market forces to reflect the cost of 'externalities' such as damage to the ozone layer, or the loss of genetic diversity through the destruction of rainforest. By the time such factors showed up on the cost projections it might be too late. The theory also assumes that the vast sections of the global economy concerned with military spending or probing space are subject to ordinary market forces and that the 'opportunity costs' of such priorities are measurable and will eventually be reflected in changing policies.

Even if market forces can be relied on ultimately to force a sustainable society upon us (and to some extent, they will), there are good reasons to meet our economic destiny as far along the road as we can. There will be less eventual repair work to be done and the social casualties of 'business as usual' are worth avoiding in any case.

What strategies are there now, then, for people who see a need for radical change in order to achieve a sustainable world, but are aware that there are, after all, no shortcuts or fixes and, worse, that there are legitimate conflicts of interest between classes, cultures and nations which must be resolved if this state is to be achieved?

The context is clearly global and the issues are moral, to do with distributional and generational justice. Morality in any context is something which develops as a guide to managing the reconciliation of individual and collective self-interest. In the intellectually and geographically insular world of early industrial Britain, it was possible for political leaders, making selective use of Adam Smith, to argue that this reconciliation was automatic, and the prospect of unlimited and continuous economic growth made his assumption entirely plausible. The best servant of the community was the individual who succeeded most in the pursuit of his own self-interest within that context. But unlike some modern economists, Smith himself stressed the importance of a pluralist approach to economic problems and regarded prudence, humanity, justice, generosity and public spirit as essential ingredients of economic intelligence (Sen 1986). Now, the collective with which individual interest must be reconciled is no longer the nation we live in or the class, race or sex we belong to, but a finite planet. This is not because the planet has shrunk but because our knowledge of it has grown. We know enough about it now to appreciate the interdependence of its ecological systems, economies and human societies. Prudence, which Adam Smith himself regarded as a paramount consideration, demands an intelligent response to the possession of new knowledge. The global context therefore transcends the particular moral codes which have developed to serve the needs of classes, ethnic groups, nations or religious traditions. Most of us, living by such guides, have therefore been caught on the wrong foot by the realisation that resources are finite and that seemingly innocent interventions in the global ecological system are apt to have far-reaching and harmful consequences. Self-interest can no longer be justified simply on the grounds that it sometimes leads to greater economic growth. Some kinds of growth are less harmful than others and some will not be harmful at all, but continued growth of the worst kind would mean ultimate catastrophe for everyone.

The effectiveness of different courses of action will always be a matter for individual tactical judgement, and will vary with time and context; with the political institutions and traditions of particular societies and the relative responsibilities of local, regional and national levels of government. There will always be the problem for reformers of whether to concentrate on the possibility of important large-scale victories, or the greater likelihood of small immediate gains (which can always be construed as palliatives) — whether, as moderates put it, to make a point or a difference. But there is a growing consensus now that, having survived successive impending Domesdays, the transition will not, after all, be cataclysmic, but piecemeal, fairly gradual and just as frustrating as the struggles for most of the worthwhile reforms of the past, however studded the story may be with heroic occasions.

The kind of revolution or reform which causes least damage to society is the kind which starts, as this one has done, at the level of individual consciousness. If things go well the time eventually comes when political parties or governments accept the leadership of a popular movement. The realisation, at an individual level, of the need for a new global morality, on the part of green activists in every affluent country, coincides at the time of writing with a movement of the environment from the periphery of politics to the centre. The first notable success of a 'green' political party was in Belgium in 1981 when, with 5 per cent of the vote, it won four seats in the lower house and five in the Senate. In Germany, the Green Party won seats in six *Land* parliaments between 1980 and 1982 and at the 1983 general elections won twenty-seven seats in the Bundestag with 5.6 per cent of the vote, a success which was exceeded in 1986.

In Britain, where first-past-the-post voting makes it notoriously difficult for small parties to win seats in

Parliament, the competition for the 'green vote' by the major parties became an almost vulgar scramble in 1986, with the prospect of an election in 1987. The Ecology Party, which was renamed the Green Party in 1985, also gained support and struggled to retain the radical initiative from the Labour Party. Its success in remaining more radical in green terms than Labour was due to its ability to avoid the dilemmas of defence and nuclear power with which a party likely to form a government had to compromise. 'The Green Tide Comes In' was the optimistic *Guardian* (4 December 1985) headline over an article describing the invitation of six 'representative greens' to lunch with Mrs Thatcher. Accustomed to projecting scenarios for a reformed society of the future, they were called upon to translate their demands into the budgetary, legislative and institutional specifics with which politicians are familiar. How much will it cost? Who will pay? Will it be new money or redirected old money?

Such questions have a ring of common sense about them, yet they imply a perception of the environment as an expensive addition to an already burdensome economic agenda rather than as something calling for a fundamental reappraisal of priorities and strategies. It has been easy for the existing political parties on both left and right, wearied by their depleted rhetoric and anxious to demonstrate modernity, to turn to the environment for a new stick with which to beat their opponents. The easy assumption that 'we are all environmentalists now' raises the prospect of cosmetic changes which will merely make environmental damage more politically acceptable than it has lately become. Rudolf Bahro (1984, p. 167) warns of the inevitable attempts of existing parties to 'try to bring the symptoms of the crisis of civilisation under control: Our assistance in doing this will always be welcome'.

But it now seems that the Chernobyl accident a few months after Mrs Thatcher's green lunch was the most dramatic of a series of events and developments which have spoilt such strategies. Like acid rain, the green house effect, the hole in the ozone layer and the peripatetic garbage freighters roaming the oceans of the world, Chernobyl demonstrated that environmental disaster is no respecter of national or ideological boundaries; that it was not the ideology of the Soviet state so much as its technology and the mistakes which could be made with it which was a danger to itself and its neighbours. It had an immediate effect in the United States on the popularity of nuclear energy. A poll in March 1982 showed 52 per cent in favour of the use of nuclear power to generate electricity, and 46 per cent against. In May 1986 only 29 per cent were in favour, with 69 per cent against. Two weeks later, when more detailed reports of the disaster were available, approval fell to 16 per cent and disapproval rose to 83 per cent (*Boston Sunday Globe* 24 August 1986). Nuclear power is now becoming so expensive as well as dangerous that the economic arguments for not proceeding with new uranium mining ventures are becoming the most politically useful ones. Issues such as the testing of nuclear weapons and the use of nuclear powered ships have become important in green politics as much because of the secrecy, the negation of democracy and the concentration of power which they facilitate as the direct environmental damage which they cause. The effect is to unite two powerful world-wide movements, peace and the environment.

The Chernobyl accident took place also at a time when much had been achieved at various levels in Western society to ensure that its implications were thoroughly understood. The changes which took place in the 1970s in the academic structure of university Centres of Environmental Science, for example, had now worked their way into secondary schools. Ministers for the environment were now to be found in every affluent society and in a good many poor ones. True, their relationship with their ancestral portfolios of 'planning' or 'development' is at times ambiguous, but in most cases it has been resolved in the direction of greater independence for the environment portfolio, and therefore authority. Like an auditor-general who keeps other departments honest, a good minister for the environment not only formulates legislation related directly to his or her responsibility but makes it impossible for other departments to be unaware of the environmental dimensions of defence, transport, energy or education.

Chernobyl was a reminder to everyone that the chances of a sudden and cataclysmic end to the human dominance of the planet have not diminished since the 1950s, but the political experience of three decades has also shown that

society is unlikely to be frightened into the adoption of
drastic simple solutions. A context in which Helen
Caldicott could hold up a baby in front of a crowd of women
and tell them they had thirty days to save the world has now
become one in which the good is no longer the enemy of the
best. It is now obvious that there will be a transition to a
post-industrial society, in spite of the best efforts of governments and industry to prevent it, which means that there
are a wide variety of ways of working with the historical
grain, both collectively and individually, towards ensuring
that such a society will be sustainable. It may therefore, be
useful to consider, in this concluding chapter, the different
levels of activity at which change is taking place. First, the
level of individual ideology and action; then, the role of the
community pressure group or organisation; and finally, the
level of political society, which includes local, state and
national government.

It is easy to undervalue the power of individual activity,
but the much maligned middle-class professionals who
confine their indications of concern to walking upstairs
instead of taking lifts, and riding bicycles to work, who
attend the occasional meeting of local residents only when
new motorways threaten their property values, may not
greatly postpone the energy crisis, nor may they do much to
equalise the cost of environmental reform, but they create a
context in which more consistent and thoroughgoing greens
can become more radical without frightening the respectable. Rudolf Bahro (1986) claimed optimistically in 1982
that 'psychologically the exodus from the capitalist-
industrial system has already begun'. In the short run, this
may be wishful thinking, but the opportunity for society to
meet its destiny half way is there. On a domestic level many
people with small incomes in affluent countries find that
not only a psychological withdrawal from the industrial
system, but also a partial physical withdrawal from the
consumer economy is the most effective way of raising their
real standard of living. Genuine self-sufficiency proved as
elusive for the teepee-dwellers of Maine in the 1970s as it
did for the communards of Nimbin, Australia, or Machynlleth, Wales, but the owner-built houses of timber, mud
brick and re-cycled stone, the vegetable gardens, the home-
made bread in the ovens of suburbia testify to the survival

and re-growth of an inherently subversive ideology in the minds of the Chernobyl generation. It tends to accelerate the growth of unemployment in the industrial sector by lowering the demand for mass-produced products, but it increases the 'useful unemployment' of those engaged in the exchange of goods and services on a local or neighbourhood basis to the confusion of both tax reformers and statisticians.

Just as covert and private changes have a cumulative social effect, so does overt action, however insignificant it may appear to be. The industrialisation of the media has, it is true, resulted in their debasement, so that they tend to concentrate, in their portrayal of human society, on the greedy, the violent, the superficial and the apathetic, but they are always open to subversion in a 'responsible' direction by making 'news' out of reformist activities and gaining the support of sympathetic journalists and media workers. Heroic actions like confronting the bulldozers or clinging to the bows of nuclear ships may seem futile, and sometimes politically ill conceived, but they operate at a fundamental level. Such strategies, if they are relentless and sustained, and used as an opportunity for public debate, have succeeded in picking away at the face of orthodoxy, one idea at a time, so that the unthinkable heresy of yesterday becomes the alternative view of today and the received wisdom of tomorrow. Since the late 1980s the 'environment' has become newly fashionable and ideological, if not political revolution has become respectable.

Like all movements for reform, the green movement develops a radical/revisionist dichotomy. On the one hand, it lives in the context of commenting 'responsibly' on environmental impact statements, seeking ways of influencing or altering the decisions made by bureaucratic and political leaders. but on the other hand, it stages sit-ins, studies and practices non-violent resistance, and in writing and public rhetoric it voices concerns which involve far-reaching changes in ways of living, which amount to personal and moral conversion.

Sometimes, however, the differences are those of taste. There are engineers who are expert and enthusiastic about the development and benefits of wind power, but not all of them wish to become vegans or live in a commune. Such differences do not have to lead to factionalism. In some ways the two parts of the movement are supportive and necessary to each other. The analysis which anthropologist Lisa Peattie (1986) makes of the Peace movement applies equally to the greens:

the movement has to play both ends: the movement aspects both pushing the politicians and building them a base of support, the politicians protecting the extremists from total repression, and using the movement dramatically like 'crowd noises off' to push politics a little way

'Crowd noises off', whether they are individual changes in ways of living, communes, protests, demonstrations or sit-ins, also ensure that the transition from industrial to post-industrial society does not have to wait until all the gloomy forecasts of the pessimists are fulfilled. Practical steps towards environmental reform achieve something directly, however small, while helping by example to undermine the ideology of industrial dominance.

It is the social climate they create and the intellectual linkages which individual actions and examples demonstrate, which enable the more spectacular gains to be made at the next level of organisational complexity, that of the pressure group and the community. This level includes environmental organisations such as Friends of the Earth, the Sierra Club and Greenpeace. The recent history of Greenpeace is a useful indicator of their potential effect. Once thought extremist by society which never doubted the wisdom of pressing on with growth regardless, Greenpeace now has a wide and sympathetic world audience and network of support. Green peace may succeed where the Luddites failed because it has succeeded in emphasising the seamlessness of the mantle which it has chosen to wear. There were those critics who applauded victories won on behalf of whales, but felt that when Greenpeace chose to protest against atomic tests on Moruroa, lines could be drawn between *legitimate* 'conservationist' protests in favour of animals and *illegitimate* 'political' ones in favour of Pacific islanders.

The French terrorist attack on the *Rainbow Warrior* in Auckland harbour on 10 July 1985 removed such nice distinctions more effectively than anything else could have done. The subsequent disclosure of the knowledge and support of the President and government of France demonstrated that Greenpeace threatens not only the philosophy of nuclear arms, but also the ideology of colonialism which enabled France to conceive of breaching the sovereignty of its old ally, New Zealand, whose soldiers had died for France in two world wars, and to exploit its control of New Caledonia and Polynesia in defence of its nuclear strategy. The result was that issues once regarded as separate, and which France wanted to be kept separate, were linked by even the most unperceptive journalists. Greenpeace had already secured the support of the New Zealand government; now, governments in Australia, Britain, the USSR, Japan and the USA recognise it as a body enjoying very widespread approval. Its ability to replace the *Rainbow Warrior* at short notice, and to continue with the planned mission to Moruroa, indicated effective international funding. Politicians who regarded Greenpeace as an extremist fringe organisation or a communist front were no longer taken seriously (Morgan and Whitaker 1986).

Though less dramatic in terms of confronting existing patterns of environmental exploitation, there are an enormous number of 'exemplary projects' which, at the same time as they confront, also build alternatives. Many examples are described in *Small is Possible*, the down-to-earth sequel to Schumacher's most famous work, by George McRobie (1981). They vary in approach and aims from the 'realist' to the extreme, just as individuals do.

Brian Martin in Australia and Rudolf Bahro in Germany express the fundamentalist approach: In Martin's (1980, pp. 12–13) words:

there are two basic approaches: building new structures and getting rid of old structures. Neither of these can be successful without the other ... At its worst, building new structures [that is local collectives, co-operatives, etc.] can be an easy way for privileged members of the middle class to opt out of the more difficult work of building alternatives which provide a direct threat to what exists.

Bahro (1986, p. 164) writes: 'There is liberation only for those who put themselves both internally and externally in a position to withdraw from Capitalism; who cease to play a role'. Frankie Ashton (1985), writing in England, is, also sceptical of such a strategy, though more sympathetic. She acknowledges the historic role played by the exemplary project and the moral stand in effecting social change, and compares the voluntary co-operatives which specialise in organic food with the Christian communities, Owenite socialists and followers of Gandhi in the past. While acknowledging that such examples have a significant effect on ideas, and have 'a legitimate place in any strategy for change', she argues that under the present circumstances of an economy dominated by all powerful multinationals, such experiments cannot alone lead the way to a restructuring of industrial society.

It is precisely because of their ostensibly innocuous character, however, that 'exemplary projects' are able to have an enormously subversive effect on the dominant view of man's place in the planetary eco-system, especially when they are successful in meeting the criteria of social utility and balanced books at a time when orthodox industrial ventures find it increasingly difficult to meet such criteria themselves. As non-renewable sources of energy decrease, nuclear accidents proliferate, and banks foreclose on an over-capitalised and overproductive but insolvent agribusiness, exemplary projects have nothing to do but survive in order to have a powerfully persuasive effect.

Many prosper as well, and it is often the most ostensibly conservative projects which therefore have the most radical effect. The famous Iona Community, for example, sought to combine the cultivation of craftsmanship and economic self-sufficiency with the spiritual regeneration of the victims of urban blight in central Glasgow and the restoration of an ancient monastery on a Hebridean island 'as a sign of the necessary integration of work and worship, of prayer and politics'. Like those inspirational medieval movements which pushed reform to the brink of heresy, it was initially viewed with some suspicion by the Church of Scotland. But just as the energies of the Franciscans and Dominicans were harnessed by the medieval papacy, controversy over the growing community within the life of the church ended when the General Assembly of the Church of Scotland unanimously resolved, in 1951, to 'bring the Iona Community within the organisation of the Church and to integrate it with the life of the Church ' (Wild Goose Publications 1985, p. 3).

The major concerns of the Iona Community are listed as 'peace and justice' 'work and the new economic order' and 'community and celebration', a framework which accords closely with that of the secular green movement. It includes and fosters such activities as the development of solar power, utilising the temperature differential between sea water and loch water as a source of heat, training in the techniques of non-violent protest and resistance, training for 'useful unemployment', art and craft education and a commitment to nuclear disarmament. Finance comes from the support of a small but world-wide society which covenants a small percentage of its disposable income to the Community, together with annual fund-raising events on the mainland and a coffee house and bookshop on the island.

Though it is true that the Iona Community is dependent on the kind of world it is trying to change, it succeeds in diverting some of the surplus produced by industrial society towards the nurture of an alternative in its midst.

The same can be said of a large number of experimental and educational organisations in Western society which do excellent business undermining the ethos of capitalism. The Centre for Alternative Technology near Machynlleth in Wales, is one of the best known. It was founded in 1974, based on a disused slate quarry, by the Society for Environmental Improvement. The organisation is ostensibly both secular and democratic, emphasising an intellectual consistency which includes concern for global inequalities, sustainable methods of food production, renewable energy, conservationist building design, whole health, and the development of local community and grassroots democracy. It does not appear to have a specifically socialist vision of the future and is opposed to exaggerations of central power of any persuasion.

With over 50,000 visitors a year, the community has become a major tourist attraction, probably the largest in mid-Wales. It had in 1987 about thirty members, all on the same wage, with an allowance for children. Ten acres are farmed using organic methods and human wastes which are composted with bracken. The tasteful public lavatories, built of slate waste, enable visitors to contribute. The quarry site encloses demonstrations of organic gardening, small-scale water power and wind power, which generates all the electricity used on the site, solar heating, energy-efficient housing, a restored, energy-efficient railway, a fish farm, biogas unit, vegetarian restaurant and bookshop. The Centre also runs short residential courses and a volunteer programme for people who want to work there.

The effect of a visit is to underline the linkages between technology and society and to undermine the belief that technology is politically neutral. As with Iona, however, it is not self-sufficiency which is actually demonstrated but the appeal of a more intelligent way of living than that to which most visitors are accustomed. It is this which keeps such communities out of the red, combined with their entrepreneurial flair and their honesty. A member makes the modest claim that 'The Centre doesn't set itself up as some kind of Utopia, just to point out some positive, sustainable ways of living gently on our planet. Despite our weaknesses, people seem to like that' (Brown 1986, p. 13).

In the Suffolk village of Needham Market the Organic Farmers and Growers Company is owned and controlled by farmers practising biological husbandry. It has over 300 members throughout the United Kingdom and is run as a commercial concern to help the members make a profit. By doing so, they also successfully demonstrate how a more sustainable agriculture can be developed in the future. It was formed in 1975 not only to market organically grown produce, but also to supply farmers who are prepared to follow biological methods with services of the kind supplied by the Ministry of Agriculture and the chemical pesticide and fertiliser trade to the conventional farmer. It insists on rigid standards, including the stipulation that soil has to be free from chemical inputs for two years before its produce qualifies as 'organic'. This regulation has now been adopted by the International Federation of Organic Agricultural Movements. David Stickland (1986) Managing Director of Organic Farmers and Growers, says that 'as we export a lot of produce to Europe and Scandinavia we need to have our standard readily accepted in those continents'.

Organic farming produces a lower yield of grain per acre than conventional chemical farming, but costs are lower and prices are higher, leading to a higher profit per acre. This provides ammunition for both sides of the radical–*realo* debate. Radicals can point out that such strategies are linked to the welfare of a discerning but affluent clientele, while the poor are condemned to continued dependence on the inferior if not harmful products of the industrial system. But it enables the Organic Farmers and Growers to use the gentle and persuasive logic of self-interest to bring about change. As at the beginning of the industrial revolution, those seeking to achieve an easy transition to a post-industrial society can argue now that doing well is also doing good. For the organic farmer, for example, not only will short-term profits be higher, but his soil will be improving all the time while the conventional farmer's soil will be deteriorating to a varying degree and as he will need increasing quantities of ever more costly chemicals his future is bleak. 'One day' writes confident David Stickland, 'The E.E.C. will involve itself in organic farming, and we wish to be in the right position when that happens', (ibid).

In Australia, as in much of North America, conditions are rapidly becoming opportune for a radical restructuring of agriculture in a more sustainable direction. Having lent ever larger sums of money in recent decades to wheat farmers in order to promote energy and capital-intensive methods of production, banks now foreclose increasingly on bankrupt properties. Banks in turn find that foreclosure is a futile exercise. Owners have been encouraged to reduce the value of their equity by such practices as overclearing of natural vegetation, which leads to a large amount of soil blowing away and the increasing salinity of the soil which remains. The climate for the serious consideration of radical alternatives is thus increasingly favourable.

In 1989 a renewal of the endemic rural crisis in the South Australian wheat belt coincided with a hike in interest rates as Treasurer Paul Keating, infatuated with economic metaphors derived from eighteenth-century physics, strove to 'cool down' an 'overheated' economy. The result was an unexpected alliance between a farming community ripe for new thinking, and its traditional political opponents, the Australian Conservation Foundation. This coincided with the successful establishment of the National Association for Sustainable Agriculture (NASA), which, like its British Counterpart, has established national standards for soil purity. Organic growers report a domestic demand for their produce which they cannot meet even when they raise their prices. Bank managers would not have to be especially bright to adopt a policy of linking loan extension and restructuring to a commitment to a reduction in chemical inputs, tree planting as a means of retaining soil and controlling salinity, and business planning on, say, a ten-year rather than an annual basis.

Methods of manufacturing as well as farming which may be relatively labour-intensive, but which have low capital costs, which provide employment as an alternative to welfare payments and which help correct imbalances between town and country, or between old decaying industrial areas and growing new ones and which utilise existing skills, should commend themselves to the leaders of ailing economies regardless of their political persuasion. Exemplary projects serve to illustrate their utility in the circumstances of late industrial society.

The Apprenticeshop, for example, in the small coastal town of Rockland, Maine, provides two-year courses in wooden boatbuilding as the central theme of a liberal education in sustainable living. It attracts apprentices and short-term volunteers from all over the United States and many foreign countries. No fees are charged for tuition and a community life is established which is cheap and uncomplicated. Profits come from the fruits of research which creates a kind of bank of appropriate technology, made available through publication; also from sale of the shop's products, classic types of wooden boat. Like organically grown food, these find a gap well up the national income scale in a market saturated with the products of a capital-intensive industry using high-technology and petrochemical materials. Business is growing rapidly.

In South Australia, 1986 was the year in which the first 150 years of colonisation by industrial society were celebrated. A dissonant but persistent theme of the five preceding years had been a community project involving the building of a wooden sailing ship, designed as a birthday present to the state for use as a training ship for young people. Denied a share of funding for 'official' projects, she became the focus of considerable public enthusiasm and volunteer participation. She also provided a large number of jobs through a government-funded job-creation scheme, and after considerable public pressure was brought to bear on the state government, was completed early in 1987 with the help of state government, local government, and private loans and gifts. She was thus the means of diverting some A$3.5 million into the alternative-technology, skill-intensive section of the economy (Young 1987).

Such exercises may be individually insignificant, but in countries which are failing to achieve the sustained economic growth which is needed to diffuse the conflicts generated by inequality, they are timely. They confront directly the problems of unemployment and alienation by the re-creation of skilfulness, the mother of independence. They also serve an educational function by recommending sustainable ways of doing things and disarming opposition on both left and right.

At the other extreme in scale, modern China, home of a quarter of the world's population, provides an example of the social and political utility of small-scale industry in a planned economy. The first Five Year Plan (1953–7) began by following the Russian model in emphasising central control and the development of heavy industry. It ran into appalling problems of supply, and created shortages of food and tensions between rural and urban areas which could not be contained. The low priority assigned to local industry proved so unpopular that the development programme and the very success of the industrialisation effort itself were threatened.

From 1957 onwards, as the policies which steered development in the direction of heavy industry and central control were abandoned in favour of the 'Great Leap Forward' of 1959 and 'walking on two legs', the growth of rural industry, small-scale and cottage manufacturing has been of increasing relative national importance. Small industry under local management has been encouraged in order to provide an efficient and flexible service to agriculture, but it has had the beneficial side-effects of utilising resources and materials which cannot be used by large-scale industry, of maintaining the viability of local communities, reducing the proportion of national resources expended in transport, and of being relatively gentle to the environment.

Since the end of the Cultural Revolution and the development of a free-enterprise sub-culture within the communist state, this long-term trend has continued, with small business, much of it in private hands, now the fastest-growing section of the economy. As most small businesses in China are labour-intensive, low in their use of energy and cater for local markets, they minimise environmental damage while equalising economic opportunity (Riskin 1987; Sigursdon 1977; Young 1989).

There is, therefore, reason to expect a movement in many parts of the world in the last decades of this century, at different times, and taking advantage of different kinds of opportunities, away from the capital-intensive large-scale production which technology has made possible towards the smaller-scale kinds of creativity which history has made both environmentally sensible and economically profitable.

But such a movement will need political support at all levels, since it is a disturbing prospect to both left and right at present. 'Dry' economists pay lip-service to the value of small business as a generator of employment, but cannot yet bring themselves to abandon a belief in very large transnational and multinational companies as the most effective generators of wealth, which they regard as even more important. Nor can they abandon the foreign policies, the international arms trade and the nuclear defence policies which are the philosophical corollaries of their economic beliefs.

As the environment is increasingly seen as an integral part of politics rather than as a peripheral issue, the most effective arena for the achievement of a sustainable society will move from the path of the bulldozers to the third arena of action, the ballot box and the legislature, but political circumstances demand different strategic priorities in different contexts.

As in the ideological transition of two centuries ago, which foreshadowed the birth of industrial society, the United States has been the source of many of the most cogent ideas in the context of that society's impending transformation. And yet, because of its diversity and size, because of the non-ideological basis of its party system and the sheer wealth required to make a political impact at a national level, the American green movement has yet to achieve significant representation in Congress. This does not reflect the level of national concern about the environment. Even by 1975, 5.5 million people contributed financially to nineteen leading national organisations, and perhaps another 20 million belonged to 40,000 local groups which were formed in the wake of the first environmental crisis of the late 1960s to fight for the conservation and preservation of natural resources, local amenities, safe energy sources, safe working conditions and consumer products (Faber and O'Connor 1989). This greatly expanded the influence of the well-established organisations such as the Audubon Society, originally formed by those New England women who resolved to eschew the practice of using birds' feathers for decorative purposes, and the California-based Sierra Club.

Earth Day in 1970 symbolised the fusion of traditional and essentially elitist preservationism and conservationism with the new middle-class, student and urban environmental concerns to form the modern American environmental movement. But the conservative and predominantly middle-class nature of the movement fostered the belief that the lobby group was the most effective political tool, and that piecemeal legislation which did not challenge basic political beliefs was the most effective way of protecting the environment.

There was much to be proud of in its achievements. By the time of the 1973 oil crisis it could point to the National Environmental Policy Act (1969), the Environmental Quality Improvement Act (1970), the Clean Air Act (1970) and the Federal Water Pollution Control Act (1972) as the result of its activities. One result of this success was that environmentalism became professionalised and institutionalised. Governments at all levels could choose to deal only with recognised representatives who in turn were able to point to their success as justification for their leadership.

But behind this success lay the category mistake which equated the national effort required to, say, put a man on the moon, with that required to 'clean up the environment'. As Mark Sagoff (1988, p. 211) points out:

Americans had a rough idea of what would be necessary to beat the Russians to the moon. The costs were reasonable; the technology available; the political forces in place. When the United States declared a 'war against pollution' in the 1960s, however, no one knew exactly what would be required to win.

The result was that in the period following the oil crisis, the middle-class basis of the movement and the faith it placed in lobbying within the system rather than seeking to change it became a handicap. It reached the limits of success within the dominant paradigm and invited a backlash from both Left and Right.

The new legislation of the 1970s meant that the United States had to spend about $271 billion between 1972 and 1979 on pollution abatement and about $518 billion between 1979 and 1988 (Faber and O'Connor 1989). So it was easy for the left to blame the environmental movement for the increasing economic problems, which the country faced as it lost markets in East Asia to Japanese competition, and for the problems of stagflation and unemployment.

On the right, President Reagan was rightly seen by environmentalists as a major threat to their cause. He could not easily undo the protective legislative achievements of the early 1970s but he could and did devise administrative strategies to circumvent them. He reduced the workforce of the Environmental Protection Agency (EPA) by 25 per cent and slashed its budget. Direct regulation of industrial conduct was increasingly replaced by what were seen as cost-effective reforms such as pollution taxes or markets for pollution rights. EPA officials were allegedly bribed by offending companies. At the same time, the public perception outside the movement was that at least environmental problems were still being tackled. The membership of some of the leading environmental organisations actually declined at a time when the need for environmental protection increased (Faber and O'Connor 1989).

This, and the onset of the second crisis of the environment triggered a general split in the environmental movement between those who continued to favour the strategies of consensus, compromise and professionalisation and those who pointed to the way in which small gains could be so easily undone, and favoured the longer course of direct action and grassroots politics.

It was not until 1987 that a resurgence became generally apparent, largely in response to a new sense of crisis, but also due to the example of Europe, and the German Green Party in particular. Charlene Spretnak and Fritjof Capra's book *Green Politics: The Global Promise* (1985) was a call to Americans for action, based this time on a philosophical foundation which went beyond the essentially anthropocentric assumptions of reform environmentalism to a recognition of the place of the human species as a dependent member of the biotic community. Murray Bookchin's analysis (cited in Porritt 1988, p. 207) of the new movements which have come together in response to the new crisis is as follows:

What I now detect is a coming together of the New Left of the 1960s counter-culture [now presumably in their 50s] the mid 1960s Social Ecology movement, the 'Earth Day' environmentalists of the 1970s, the Feminist movement and the anti nuclear movement.

One of the strategies proposed by Spretnak and Capra for American readers was the establishment of a 'Green Network'. This was clearly linked with the theme of American national origins, of making a fresh start, as at the time of renewal at the end of the eighteenth century by the establishment of 'Committees of Correspondence', echoing the social movement of 1774 which underpinned the revolutionary action of 1776. 'What we are trying to do', says Bookchin, 'is to redeem certain aspects of the American Dream' (op cit., p. 210). There are, after all, several versions of this. There is the frontier dream of pioneer individualism, epitomised by John Wayne. Then there is the immigrant dream, of a land of opportunity in which anyone can become a millionaire, or President for that matter. But deeper than these in the national unconscious is the Puritan dream, derived in turn from the common source which also inspired the Diggers, Ranters and Quakers of England's century of revolution. In contrast to the concepts of the nation-state, Cromwell's military commonwealth, and the intellectual climate of Baconian science from which the Pilgrims escaped, it emphasised decentralisation, community, self-sufficiency and mutual aid.

The first National Education Conference of the Green Committees of Correspondence was held at Amherst, Massachusetts, in July 1987. Though reported as a somewhat fractious occasion, it was followed a month later by the formulation of SPAKA, a participatory process designed to formulate Strategy and Policy Approaches in Key Areas on a national basis (Adair and Rensenbrink 1989). Local committees were asked to prepare statements covering the ten themes of ecological wisdom, grass roots democracy,

personal and social responsibility, non-violence, decentralisation, community-based economics, post-patriarchal
values, respect for diversity, global responsibility, and
sustainability Three hundred representatives of some two
hundred groups met the following year in Eugene, Oregon, at
the summer solstice, 21–25 June. The next national green
gathering is planned for Boulder, Colorado in September
1990.

The central theme, as emphasised in the twin organs of
the movement, *Green Letter: Greener Times* is the virtue of
diversity and decentralisation:

In Eugene we created an atmosphere that respected different
political perspectives. Factionalising did not take hold as it had in
Amherst; instead there was a free dialogue so that all sides learned
from one another, and we grew to respect each other rather than
dismissing other points of view as simply not 'truly green' (Adair
and Resenbrink 1989, p. 5).

If a green political party does develop in the United States,
it will thus be one which is firmly embedded in a social
movement. Consensus was reached in the American context, and in view of the enormous expense of the electoral
process, that lone candidates with no grassroots base risked
marginalisation or co-option. There is no intention of running national candidates till, perhaps, 1992, and no discussion of a presidential candidate until, at the earliest, 1996.
Meanwhile, the major effort will be concentrated at local
county and municipal level, in keeping with the bioregional
approach of many of the committees. From an Australian
perspective the movement seems lacking in awareness of
urban environmental problems and an international consciousness, while its tendency to lump together 'people of
colour' as a single constituency glosses over the very specific
and distinctive contribution of indigenous peoples to post
environmentalist theory. But the movement rejects the
notion of seeking to reconcile the present economic system
with the environment. It counts on building counter-
institutions within the body of industrial society and it
combines an awareness of internal problems with an enormous energy and a determination to solve them.

In contrast to the US greens, the experience of the
German Green Party illustrates the opportunities and

problems of a parliamentary system with proportional representation, preferential voting and financial arrangements which favour the electoral activities of small parties.

Since the electoral success of 1983, the issue is no longer whether to accept the compromise of principle involved in entering the parliamentary arena at all, but, once within it, whether to contemplate coalition with other parties in order to participate in government, and at what price, or to remain independent (Spretnak and Capra 1985). There are good arguments on both sides but to obtain redistribution of money from, say the nuclear programme or the arms industry to the encouragement of self-sufficient communities even on an experimental basis means helping, for the time being, to keep an existing party or coalition in power, and risks co-option and the dissipation of political energy. Some prefer to achieve what they can without hoping for government support in order to avoid such compromises.

Divided as it is over the significance of parliamentary activity as against direct action or grassroots organisation, the Green Party insists on safeguards against the reformist effect which deliberative politics inevitably has on those who take part in it. Members of the Bundestag are debarred from holding office in the party; and party offices used to rotate every two years, though this is no longer the case.

While these measures guard against the classic transition in identity from representative of a constituency to member of a deliberative legislature, they are symptomatic of a party which does not expect to govern in the foreseeable future. Its rhetoric indicates the assumption that power will come eventually and suddenly, after sweeping change or the collapse of present structures followed by a combination of overwhelming grassroots support and the winning of an independent parliamentary majority. Such expectations provide hope for those opposed to compromise:

It is precisely our pressure from an Archimedean point outside the previous world of ideas that can effect a change in the political landscape. The more unavailable we are for the S.D.P., the stronger in its inner party power play will be the wing that is turned towards us' (Bahro 1986, pp. 47–8).

Historically, the choice between the alternatives of ideological purity and small gains has rarely been absolute;

the most lasting reforms require some element of continuity between the old state of society and that which replaces it. For this, reformism and radicalism need to develop a complimentary relationship. First-past-the-post systems of election, which do not favour the growth of new parties, as in Britain, encourage this relationship to develop within existing parties rather than between parties in Parliament. The major parties began by proposing cosmetic changes in policy to prevent the 'green vote' falling into the hands of their opponents, a strategy which tends to get them into intellectual trouble as greens succeed in developing their ideological consistency. One Conservative pamphlet, for example, explains that party members and voters are not sufficiently involved in 'environmental organisations.'

The tendency is for some of these, both to their own detriment and that of conservation generally, to come under left-wing influence. It is most important that these bodies remain a-political in party terms . . . the Conservative Party must therefore play its role through its supporters, in 'holding the ring' (Patterson 1984, p. 19).

Other Tories, more sincere in their desire for reform, are nevertheless apprehensive about the appeal of environmentalism to the vast section of the middle class represented by such societies as those interested in hunting, fishing or collecting butterflies. Kenneth Carlisle (1984) cites the 1 million members of the National Trust and the half million members of the Royal Society for the Protection of Birds as a vital constituency. Andrew Sullivan (1985) seeks to retain the support of the millions of television viewers who watch *All Creatures Great and Small*, based on James Herriot's stories about a country vet, or who delight in the humorous snobbery of *To the Manor Born*, whose hostility could be aroused by the image of a government identified with property development, pollution and economic growth at the expense of the environment: 'the Green issue will not go away. The correct and healthy Tory reaction is to expropriate it' (Sullivan 1985, p. 44).

The problem for conservatives, however, is how green can they get and still remain blue? The fact of a green movement, not just individual concern about the environment, and the increasing sophistication of that movement, makes it difficult for those intent on 'holding the ring' not to be confronted by their own naivete as they examine the linkages between such things as industrial growth, acid rain, fish deaths in Scandinavian lakes, or EC agricultural policies and the removal of English hedgerows. Green organisations should therefore welcome Conservatives to their ranks as potential converts. If they remain members of the Conservative Party, and that party remains in power, at least small gains may be made. The pressure of the environmental movement, combined with the logic of economy and efficiency, is likely at some stage to reverse such policies as favouring road over rail transport in any case. Andrew Sullivan (1985, p. 41) argues that 'A stronger Tory commitment to public transport may, in the present context, be heresy, but it is in line with many popular sentiments, in touch with the particular demands of the English environment, and *even economically worthwhile*. Such reasons make even heresy attractive'.

British centre parties have not succeeded in thinking the implications of green politics through as well as might have been expected from their middle-class support. The Social Democratic Party (SDP) approach is a selective one, aiming at what it calls 'green growth', with a 'balance' struck between growth and, *where necessary*, which is presumably not all the time, 'environmental protection'. The criteria for such necessity is not explicit and it is significant that environmental policy is presented in terms of conservation. An aspect of class in Britain is that many SDP voters have an emotional and cultural commitment to France, and are consequently enthusiastic about the idea of Britain as part of Europe. They have faith in the EC as a means to the achievement of wider international co-operation and accept French arguments in favour of an 'independent European nuclear deterrent' as a balance to the nuclear forces of the USA and the Soviet Union. The cessation of atomic testing on a unilateral basis by the latter therefore placed the SDP in a difficult position and further superpower agreement on disarmament might prove embarrassing. Its plans for nuclear collaboration with France, and therefore with French nuclear testing in the South Pacific, means it must support the maintenance of a French military presence and the suppression of Kanak independence in New Caledonia. Such commitments sit uncomfortably with a party which calls itself 'Democratic'.

The implications of foreign and economic policy for the environment were a major cause of tension between the Liberal Party and the S.D.P. which formed an uneasy alliance from 1982 until 1987. In that year, the S.D.P. split, with the majority joining the Liberals to form a new Liberal Democratic Party. The S.D.P. rump, led by Dr David Owen, has remained wedded to nuclear energy as well as to nuclear weapons. The Liberal Democrats, however, like their Liberal predecessors, are in favour of phasing out nuclear power, and believe that present living standards can be maintained with less than half the current consumption of fossil fuels. They see the arms race as a major environmental threat as well as a threat to immediate safety. Hutchinson (1986, p. 10) pointed out that, 'We now live with the disgraceful reality that world military expenditures each year not only exceed the combined income of the poorest half of the world, but are growing faster than the world economy as a whole.'

In countries like Britain and France, competition for the huge, but hitherto frustrated green vote, in the context of the revolutionary events in Eastern Europe at the end of 1989 could lead to a radical re-alignment of political forces. Parties like the Liberal Democrats now have a conjunction of ideological interests with both green parties and with the powerful movements from Scotland to Azerbaijan which are campaigning for regional or national autonomy, and which are often fuelled by environmental as well as cultural concerns. Old, normally socialist parties, moving steadily to the right in search of central power, could find themselves much closer to conservative governments committed to economic growth and to the maintenance of cold war postures than to such an opposition.

In Australia the Labour Party has also moved to the right and has become the most successful champion of big business in the country's history. This has placed the Democrats, once a 'centre' party, well to the left of both main parties, and it is now identifying itself successfully as Australia's green party. The Liberal Party confronts the forlorn task of trying to occupy a position both popular, and to the right of Labour, by clutching at such straws as a racially restrictive immigration policy, more 'law and order' and a foreign policy less sympathetic to the Soviet Union, while at the same time becoming the unlikely champions of the widows and orphans of society, poor families and pensioners.

A 'green faction' has now developed within the Labour Party. It has little chance of gaining dominance of a party now openly patronised by the Bond Corporation, on the one hand, and the trade union movement, an almost equally conservative force, on the other, and will either have to compromise its principles or break away. If it does, there are a large and growing number of independent movements, some of them green, with whom green Labour and an increasingly popular Australian Democrat Party might join to form an effective green coalition. In the present context it would have no opponents to the left. In Tasmania, five green independent Members of Parliament held the balance of power after the May 1989 state elections. They have made an alliance but not a coalition with the state Labour Party to allow it to govern in return for the protection of large additional areas of forest, and a range of green initiatives on social and employment policy. While Labour, on a federal level, is prepared to compromise, for the sake of power, over local issues such as logging in National Estate areas, and will always respond to conservationist pressure at election time for the preservation of wilderness, it will not be able to abandon the philosophical basis of the dominant paradigm which it shares with its Liberal opponents. Nor will it lightly abandon the 'economic rationalism' which brought it electoral success in 1983. It thus promotes the uranium industry and supports existing farming practices just as enthusiastically as its opponents from the rural electorates. It will therefore be an unlikely vehicle for the kind of radical reforms which will be necessary in the decades to come. The British Labour Party, too, is vulnerable to the charge that though theoretically it stands against capitalism, and identifies capitalism as the cause of the environmental crisis, it does not challenge capitalism in practice. In the area of environmental politics it concerns itself with issues like danger and pollution rather than the causes of such problems while modern Unionism conforms increasingly to the rules of capitalist enterprise, seeking the best prices for the labour commodity which it controls and engaging in collusive bargaining at the expense of both environment and consumer.

But there is now a powerful movement within both the trade union movement and the Labour Party which is working for an alliance with the green movement. Proponents argue that pressure-group environmental activity, independent of party allegiance, is important, but that it is also important to have in power a political party which is the one most likely to be receptive to the pressures which are set up by the green movement. In view of the very strong lobbies in favour of more motorways, armaments, rainforest destruction, and so on, which expect concessions from Conservative governments, they argue that Labour is the environments' only real hope. They believe that in view of the potential within the Labour movement for the revival of the anarchist rather than a centralist socialist tradition, it provides the best defence the environment is likely to get (Pepper 1985). At the same time there is scope for work within the Labour movement in order to strengthen the kind of socialism the greens want to see.

The British Green Party changed its name from the 'Ecology Party' at its 1985 Dover conference. Its 1986 manifesto, 'Politics for Life', starts with a quotation from Schumacher which implies resignation to minority status, and encapsulates the justification for voting for such a party, which many see as 'throwing away a vote':

We must do what we conceive to be the right thing and not bother our heads or burden our souls with whether we're going to be successful. Because if we don't do the right thing, we'll just be part of the disease and not part of the cure (Green Party 1986).

Policies therefore tend to express long-term ideals rather than immediate improvements. The party aims, for example, to 'Establish a programme for increasing British agricultural self-sufficiency to produce 90 per cent of the food required, based on mixed small farming.' What happens to those successful East Anglian Eurofarmers who want to go on prospering the way they are now we are not told. The party favours not 'sustainable growth' after the fashion of the Labour Party, but no growth at all. It favours decentralisation, and, in power, would take Britain out of both the EC and NATO. It favours regional governments for Wales and

Scotland and would aim for a long-term reduction in population. It would build no more motorways or airports and would revive railways, canals and coastal shipping.

But just as existing political parties are now forced to include the environment as a central part of their thinking, the Green Party has been forced to think about issues like defence, and has not found it congenial. In the long run, of course, nuclear defence would be abandoned, attempts would be made to achieve a non-aligned, neutral Europe and current civil defence programmes would be scrapped in favour of training in non-violent resistance to invasion. What is missing are the intermediate strategies in the event of failure to achieve a non-aligned Europe and details for the redeployment of NATO forces. Land reform is a major means of achieving social change. It is proposed that the speculative value of land should be removed by the simple expedient of imposing a charge equal to its full ground rent so that the benefits of possession would derive solely from use. This might have an equalising effect, but when the 'single taxer' Henry George originally introduced the idea in the 1880s it was also intended to insure that land achieved maximum productivity. It might do little to halt the destruction of the soil and, as an isolated measure, might accelerate it.

The Green Party is thus likely to be effective for some time to come as a party of opposition, and the same is true of similar parties in Europe. Meanwhile, it is worthwhile for all those concerned with the maintenance of a sustainable world to plan for the immediate as well as the distant future. Some may achieve most by supporting a green party or organisation in its long struggle for power. Others can also ensure that after green governments come to power, and when they lose office to their rivals, as in a democratic and decentralised society they must, the earth itself will not be threatened whatever parties or interests come to dominate politics. Environmental wisdom should become as enshrined in the politics of all parties as honesty is supposed to be now. One long-term way of ensuring compatibility between an 'Ecotopia' of the future and the political freedoms which have made the development of green politics possible is to ensure that within each existing political party there is a substantial constituency which will promote ecologically sound policies. Such are the aims of such organisations as the Australian Conservation Foundation and the Green Alliance, a small London-based organisation funded by the private subscriptions of a relatively small number of individuals. It exists to collect and disseminate information and to act as a green lobby whatever kind of government is in power.

Just as the search for consistency has widened the area of green commitment from conservation to disarmament, it increasingly includes questions of international equality, of aid and trade. Green radicals like Bahro in West Germany, Michael Redclift in Britain and David Brower in the United States have all followed the logic of this train of thought to its conclusion and added another plank to the green platform. 'Instead of starting new export drives', says Bahro (1986, p. 37) 'and thereby exporting some of our unemployment to other countries, we want needs-oriented work in economic and living zones which are as decentralised as possible'. Michael Redclift uses the example of Guatemala to illustrate the inequalities which are produced by present patterns of world economic growth. Seventy-five per cent of the money earned from cotton exports, for example, goes to pay for the pesticides, spray planes, tractors and other imported inputs needed to produce the cotton. Local manufacturing, and export of clothing instead of the raw material would mean that only 1 per cent of the land now cultivated would be needed to earn the same amount of foreign exchange. The same kind of calculation applies to the imported Japanese machinery used to destroy the Tasmanian rainforest to supply the Japanese packaging industry with woodchips. A withdrawal of labour and capital from American clothing factories and Japanese machine shops for redeployment in regional self-sufficient forms of production would have parallel effects in poor regions like Guatemala and Tasmania.

The Philippines, which used to be cited as a model of development success, along with South Korea, Taiwan and Singapore, now provide the clearest example, both environmentally and politically, of what not to do. This is sometimes assumed to have been the result of unusual corruption on the part of the Marcos regime, but the attempt to manipulate a poor country in the direction of both industrialisation and militarisation is always likely to lead to a situation in which violence and corruption become the means of government. As Jonathon Porritt (1984) puts it:

Boosting exports usually means pushing peasants off their land to make it available for cash crops for exports to pampered consumers in the west. Borrowing more money increases the debt burden, the interest on which goes to western banks ... the inevitable result is poverty, opposition and conflict, at which stage, arms come in very handy to suppress dissent in the name of stability and 'freedom'. And so the cycle goes on.

Conversely, land reforms and proposals of the Aquino government to redress social injustice will also have the effect of arresting the process of environmental destruction and will create the possibility of a sustainable economy some time in the future. The problem is to overcome the inertia of past military investment and the politicisation of the military forces in order to achieve reform with sufficient rapidity to forestall further revolution from the left.

With such an example in front of them, poor countries which have succeeded so far in remaining on the periphery of the global economy will be well advised to refuse such inducements as commercial television, however much their elites may demand them. Like transistor radios, television sets will be used primarily to create new wants, only to be satisfied by the abandonment of subsistence production and local trade and migration to the already crowded and dangerous squatter settlements on the outskirts of the major cities. They should refuse the advice of economists who will tell them that they must encourage overseas investors to shift modern factories into their cities to take advantage of the cheap labour to be found there. Above all, they should refuse military aid, especially if it involves the establishment of military bases and includes military advisers who will need to be housed in air-conditioned premises, fed with imported food, armed by the expenditure of hard-earned foreign exchange, and provided with prostitutes drawn from the food-growing female section of the population.

Countries whose path of development has led to extremes of internal inequality, such as the Vietnam of the 1960s, Nicaragua or the Philippines, have inevitably found themselves locked into internationally unequal distributions of wealth and power and have sought revolutionary remedies. It is, therefore, not surprising that the more equal ones have been able to achieve the liberty to act in ways which increase international equality and also tend to the benefit of the planet. The record of the more equal wealthy societies like New Zealand and Sweden shows that if the moral judgement of majorities is not distorted by envy of their own rich, well-intentioned governments are free to spend more of their national incomes on aid and to administer their aid programmes with greater sensitivity to the needs of the recipients. New Zealand, for example, does not make the provision of a market for New Zealand produce an implicit or explicit condition of assistance to neighbouring Pacific island states. It is not a coincidence that the government of New Zealand should be the first sovereign power to provide unequivocal support for Greenpeace, and to ban the visits of nuclear ships.

Choice of individual action on behalf of the environment and in the hope of achieving a sustainable world will vary in different countries and circumstances but is likely to be effective to the extent that its consequences within a planetary as well as a local context are considered. The Newtonian perception of the earth as a kind of machine and nature as a utility is the philosophical expression of the economics associated with Adam Smith in which self-interest becomes a virtue in the secular context of unlimited growth. Recognition that growth cannot be unlimited leads logically to the conclusion that, as humanity has found for most of its history, self-interest and the interests of mankind are not in automatic harmony, and that ultimately, the conditions for reform, whatever their cultural context, have a moral and spiritual dimension. The choice between starting new green parties or greening existing parties, between the 'moral suasion' of writing to newspapers and the direct actions of hanging on to the bows of nuclear ships or spiking trees, will continue to be determined by individual temperament and opportunity, but the ways in which such disparate actions are mutually supportive morally are ultimately more important than ways in which they embarrass each other politically.

Most pressing of the moral problems of late industrial society is that instead of providing a growing demand for labour, a shrinking labour force of increasing expertise (which is not the same thing as skill) will produce more. As with agriculture in the 1960s, industry in the 1990s and beyond will be able to produce a mountain of goods for those who have jobs, and can afford them, with fewer producers. The consequence will be the growth of what some writers call an 'underclass' of unemployed and poorly paid part-time workers.

In Britain and America, and now in Japan, Australia and New Zealand, the underclass is already numerous. It consists of migrant workers, living in inner-city areas, often in single-parent families, with a low rate of literacy and a high rate of drug addiction and associated criminality. They are deemed unlikely to form the vanguard of a revolution because they are not necessary to the economy and some politicians and economists say their existence is the price modern society must pay for its success.

But the cost to the society from which this large section of its youth has been removed is enormous. No-go areas of the once great and civilised cities of the world expand as a result of an effective denial of citizenship to whole categories of people. 'Normal' society closes its ranks, fastens its deadlocks, talks of 'law and order' and, with the aid of the unions, the police and the bureaucracy, distances itself from them. To the extent that the underclass is composed of young people it is distancing itself also from its future. The AIDS virus may turn out to be only one of the ways in which, to quote Ralf Dahrendorf (1987), 'Defining entire categories out will in the end rebound on those who believe it gives them safety and comfort'.

Suggested remedies have come mainly from the Left. In 1984 German metalworkers, for example, went on strike, not for more money, but for shorter hours. Australian unionists have achieved similar results with less trouble in one industry after another. Jobs can be shared, but it is much easier to share the kinds of job which are creative and stimulating like teaching than it is to share dull jobs on the assembly line, and because part-time work is always less secure than full-time work it is unlikely to be politically acceptable as a general remedy for unemployment unless it is coupled to the idea of a basic guaranteed income regardless of employment. All the usual weapons of the

Right, and arguments about dependence and lack of incentive to do any work at all, would be brought into play to defeat such schemes in most capitalist societies.

Green politics provides a way out of this impasse and into a sustainable and equitable future without sacrificing the advantages of new technology. Work rationing need not take the form of steadily reducing the hours worked at polluting, resource-exhausting productive processes. It could take the form instead of *advancing* (rather than returning) to methods of production which are gentle to the environment and more sustainable. The present response to unemployment by benign governments is to spend money creating jobs, within the existing industrial framework. Declining real job opportunities mean that they have to be temporary to make sure that as many people as possible get a turn at a boring job. This both returns people temporarily to the ranks of the consumers and improves the 'job-creation' statistics.

An alternative strategy would be to work through community organisations and local governments to divert funds solely into projects which develop the kinds of skill which enable people to sustain and employ themselves. The aim should be to provide not just temporary relief from a place in the dole queue but educational and practical experience of a consciously creative and therefore subversive nature. The criteria for funding should be the extent to which workplaces are provided at low cost and require minimum use of non-renewable forms of energy. The use of skilfulness and of recycled or renewable materials should be encouraged, together with the best and most advanced technology available for these purposes. The object would be to remove as many people as possible from as much as possible of the consumer economy. The paradoxical effect would be to regain them as members of society. GNP might suffer for a while, until ways of measuring it were reformed, but welfare payments would be reduced, which would reduce taxation. Since such a strategy would provide increasing scope for individuality, and because it would stress liberty, while achieving a greater measure of equality, it would win support from the Right as well as the Left. Unemployment, the incurable symptom of a sick society, will respond only to some such antibody and would lead to logically compatible policies in areas such as agriculture, energy, defence, aid, education and medicine.

The somewhat puritanical perception of Ivan Illich, that modernised poverty is the consequence of the abandonment of personal autonomy, of being 'plugged into market relations', is perhaps the first stage in the revival of an ethic which is compatible with a sustainable society.

Subsequent stages will vary according to place, culture and religious tradition. The Islamic revival, for example, is to a large extent a rejection of Western materialism and the corruption which is seen as its consequence, and the cultural revivals elsewhere in the post-colonial world have similar ingredients. If sustainable post-industrial societies are to be achieved, they will probably not arise from the ashes of a nuclear holocaust or from the survivors of a revolutionary apocalypse. The best chances of survival are for those social states which will be reached in irregular fashion, as individuals and families, streets or villages, suburbs, cities and the more sensible national governments which represent them respond to the series of problems and crises which lie ahead. As fossil fuels are exhausted in the twenty-first century and as the cost of militarisation not only of the earth, but also of space, becomes unbearable, first for one superpower, then the rest, people will discover that their ability to live better as well as cheaper depends on the extent to which they can become masters rather than servants of technology and liberate themselves from the imperatives of industrial society. Politicians, probably in small countries to begin with, will sense votes in the positive encouragement of alternative economies, and if present trends continue the time will come when national leaders will be able to count on majorities to understand the consistency of programmes of reform which encompass such superficially disconnected but morally consistent purposes as restoring land to indigenous peoples, preserving rainforest, subsidising urban farms and banning nuclear warships. Large countries, hoping to be seen as leaders, and under pressure from their own huge minorities, will have to adjust to a world ready to accept responsibility for the consequences of moral as well as political independence.

Suggestions for further reading

Jonathon Porritt and David Winner's book, *The Coming of the Greens* (1988), gives a great deal of recent detail about the green movement in Britain and puts it in the context of the global green movement. Brian Tokar's *The Green Alternative* (1987) deals with green developments in the United States.

A hopeful sign of change in the relationship between rich and poor countries of the world is a recent report to the Club of Rome, by Bertrand Schneider, *The Barefoot Revolution* (1988).

The religious dimensions of green politics are discussed in Sean McDonagh, *To Care for the Earth* and Wendell Berry's *Continuous Harmony*. Appropriate technology is now a rapidly developing field. Marilyn Carr's *A.T. Reader: Theory and Practice in Appropriate Technology* (1985) is a good start.

Bibliography and references

Abram, David (1985). 'The Perceptual Implications of Gaia', *Ecologist*, 15:3, 96–103.

Adair and Rensenbrink, J. (1989) 'Strategy and Policy Approaches in Key Areas' (SPAKA). Meeting, Eugene, Oregon USA. *Greener Times* Autumn, 1989.

Ashby, E. (1978). *Reconciling Man with the Environment*, Oxford University Press.

Ashton, F. (1985). 'Green Dreams, Red Realities', NATTA Discussion Paper no. 2, Milton Keynes, November.

Attfield, R. (1983), *The Ethics of Environmental Concern*, Basil Blackwell, Oxford.

Augustine (1961). *Confessions*, translated and with an introduction by R.S. Pine-Coffin, Penguin Books, Harmondsworth.

Bacon, Francis (1627). 'New Atlantis' in Thomas Case (ed.), *The Advancement of Learning and New Atlantis*, Oxford University Press (1906) 228.

Bahro, Rudolph (1979). *The Alternative in Eastern Europe*, translated by David Fernbach, Verso, London.

Bahro, Rudolph (1986). *Building the Green Movement*, translated by Mary Tyler, Heretic Books, London.

Barbour, Ian G. (ed.) (1972). *Earth Might Be Fair*, Prentice Hall, Englewood Cliffs, NJ.

Barbour, Ian G. (ed.) (1973). *Western Man and Environmental Ethics: Attitudes Towards Nature and Technology*, Reading, Mass.

Barbour, Ian G. (1980). *Technology, Environment and Human Values*, Praeger, New York.

Bateson, G. (1980). *Mind and Nature, A Necessary Unity*, Fontana, London.

Bayliss-Smith, T.P. and Feachem, R.G. (eds) (1977) *Subsistence and Survival: Rural Ecology in the Pacific*, Academic Press, London, New York.

Bayliss-Smith *et al.* (1988) *Islands, Islanders and the World: The Colonial and Post-Colonial Experience of Eastern Fiji*,

Cambridge University Press, Cambridge, (UK), New York.

Beaglehole, J.C. (ed.) (1955–67). *Journals of Captain Cook*, 3 vols, Cambridge University Press.

Beaglehole, J.C. (ed.) (1962). *The Endeavour Journal of Sir Joseph Banks*, 2 vols, Angus and Robertson, Sydney.

Beckerman, W. (1974). *In Defence of Economic Growth*, Jonathan Cape, London.

Bellwood, P. (1978). *Man's Conquest of the Pacific*, William Collins, Auckland.

Beresford, M. (1978). 'Doomsdayers and Econuts: A Critique of the Ecology Movement', *Politics*, 12, 98–106.

Berman, M. (1981). *The Re-enchantment of the World*, Bantam Books, New York.

Berry, Wendell (1975). *Continuous Harmony*, Harvest Books, San Francisco.

Berry, W. (1987). *Home Economics*, Schumacher Society.

Birch, C. (1976). *Confronting the Future*, Penguin, Harmondsworth.

Birch, C. (1984). 'Born Again Science and Technology', Inaugural Keith Roby Memorial Lecture in Community Science, 1982, Murdoch University, W.A.

Birch, C. and Cobb, J.B. (1981). *Liberation of Life: From Cell to Community*, Cambridge University Press.

Black, J. (1970). *The Dominion of Man: The Search for Ecological Responsibility*, Edinburgh University Press.

Blainey, G. (1982). *The Triumph of the Nomads: A History of Ancient Australia*, revised ed., Macmillan, Melbourne.

Bookchin, M. (1971). *Post-Scarcity Anarchism*, Ramparts Press, Berkeley, Calif., Wildwood House, London.

Bookchin, M. (1980). *Towards an Ecological Society*, Black Rose Books, Montreal.

Bookchin, M. (1982). *The Ecology of Freedom*, Palo Alto, Calif.

Bookchin, M. (1986). *The Modern Crisis*, New Society Publishers, Philadelphia.

Bookchin, M. (1987). *The Rise of Urbanisation and the Decline of Citizenship*, Sierra Club Books, San Francisco.

Bookchin, M. (1987). 'Social Ecology versus Deep ecology'. A Challenge for the Ecology Movement', *Green Perspectives*, nos 4 and 5.

Borelli, P. (1988). 'The Ecophilosophers: A Guide to Deep Ecologists, Bioregionalists, Greens and others in pursuit of Radical Change', *The Amicus Journal*, 10:2.

Bramwell, A. (1989). *Ecology in the 20th Century, A History*. Yale University Press, New Haven, Conn.

Brekilien, Y. (1982). 'The Red Regions of France, *Ecologist*, 12:5, 217–26.

Brewer, J.S. (ed.) (1859). *Opera Fr. Baconis Hactenus inedita*, London.

Bronowski, W. (1964). *Science and Human Values*, Penguin, Harmondsworth.

Brown, L. (1981). *Building a Sustainable Society*, W.W. Norton, New York.

Brown, Paula and Buchbinder, G. (eds) (1976) *Man and Woman in the New Guinea Highlands*, American Anthropological Association, Washington D.C.

Brown, Tim (1986) *Network for Alternative Technology and Technology Assessment* (NATTA). Newsletter May/June 1986, Milton Keynes.

Bunyard, P. and Goldsmith, E. (1988). *Gaia: The Thesis, the Mechanisms and the Implications*, Wadebridge Ecological Centre, Camelford, Cornwall.

Burke, E. (1790). *Reflections on the Revolution in France* (ed.) J.G.A. Pocock (1987) Hackett Publications, Indianapolis, Indiana.

Burklin, P.W. (1985) 'The German Greens: The Post-industrial Non-established Left and the Party System', *International Political Science Review*, October.

Burch, D. (1982). 'Appropriate Technology for the Third World: Why the Will is Lacking', *Ecologist*, 12:2.

Caldwell, M. (1977). *The Wealth of Some Nations*, Zed Press, London.

Callenbach, E. (1975). *Ectopia: the notebooks and reports of William Weston*, Banyan Tree Books, California.

Callicott, J. Baird (1982). 'Hume Is/Ought Dichotomy and the Relation of Ecology to Leopold's Land Ethic', *Environmental Ethics*, 4:2, 163–74.

Callicott, J. Baird (1987). *Companion to A Sand County Almanac: Interpretive and Critical Essays*, University of Wisconsin Press, Madison.

Capra, F. (1983). *The Turning Point: Science Society and the Rising Culture*, Fontana, London.

Carlisle, K. (1984). *Conserving the Countryside: A Tory View*, Conservative Political Centre, London.

Carr, Marilyn (ed.) (1985). *The A.T. Reader: Theory and Practice in Appropriate Technology*, Intermediate Technology Publications, London.

Carson, R. (1956). *The Sea Around Us*, Penguin Books, Harmondsworth.

Carson, R. (1962). *Silent Spring*, Houghton Miflin, Boston.

Carson, R. (1964). *The Sea*, McGibbon and Kee, London.

Clarke, R. (ed.) (1975). *Notes for the Future: An Alternative History of the Past Decade*, Thames and Hudson, London.

Clarke, W. (1977). The Structure of Permanence: The Relevance of Self-subsistence Communities for World Ecosystem Management' in Bayliss-Smith and Feachem (1977).

Cole, H.S.D. et al. (1973). Thinking About the Future: A Critique of the Limits to Growth, University of Sussex Press, Falmer.

Collingwood, R.G. (1945). The Idea of Nature, Oxford University Press.

Commoner, Barry, (1972). The Closing Circle, Alfred Knopf, New York.

Commoner, Barry (1976). The Poverty of Power: Energy and the Economic Crisis, Jonathan Cape, London.

Commoner, Barry (1979). The Politics of Energy, Knopf, New York.

Cooley, M. (1980). Architect or Bee? The Human-Technology Relationship, Transnational Co-op Ltd, Sydney, Australia.

Cooley, M.J. (1983). The New Technology: Social Impacts and Human Centred Alternatives, Technology Policy Group, Open University, Milton Keynes.

Coombs, H.C. (1989). Post Scripts: the 1988 Boyer Lectures, pp. 95–107, A.B.C. Enterprises Ltd, for the Australian Broadcasting Corporation, Sydney.

Cotgrove, S. (1982). Catastrophe or Cornucopia: The Environment, Politics and the Future, John Wiley and Sons, New York.

Cowen, David, (1985). 'Religion, Health and Sickness: The Process of Conversion in Fiji', unpublished paper, University of Adelaide History Department.

Curzon of Kedleston (1919). Report of the Privy Council for Scientific and Industrial Research,Cmd 320, HMSO, London.

Crick, F. (1960). Of Molecules and Men, University Press, Washington.

Dahrendorf, R. (1987). 'The Erosion of Citizenship and its consequences for us all', New Statesman and Society 12 June 1987, pp. 12–15.

Daly, H.E. (ed.) (1973) Towards a Steady State Economy, W.H. Freeman & Co., San Francisco.

Daly, H.E. (1977). Steady State Economics: The Economics of Biophysical Equilibrium and Moral Growth, W.H. Freeman & Co., San Francisco.

Daniellson, Bengt and Marie-Thérèse (1977). Moruroa Mon Amour: The French Nuclear Tests in the Pacific. Penguin Books, Harmondsworth.

Davis, Dorothy et al. (1980). Living Together: Family Patterns and Lifestyles, a book of readings and reports. Centre for Continuing Education, Australian National University, Canberra.

Dendy, Tim (ed.) (1988). Greenhouse '88: Planning for Climate Change. Adelaide Conference Proceedings, Department of

Environment & Planning,, South Australian Government.

Descartes, R. (1975). *A Discourse on Method*, Everyman, London (first published 1637).

Devall, Bill (1980). 'The Deep Ecology Movement', *Natural Resources Journal*, 20:2.

Devall, Bill and Sessions, George (1985). *Deep Ecology: Living as if Nature Mattered*, Peregrine Smith, Salt Lake City.

Dobb, M. (1973). *Theories of Value and Distribution since Adam Smith: Ideology and Theory*. Cambridge University Press, Cambridge.

Dodge, Jim (1981). 'Living by Life: Some Bioregional Theory and Practice' *Co-Evolution Quarterly*, Winter 1981, pp. 6–12, Sausalito, California.

Dyer, K.F. and Young, John (eds) (1990). *New Directions*: Ecopolitics IV Proceedings, University of Adelaide 1989.

Easlea, Brian (1973). *Liberation and the Aims of Science: An Essay on the Obstacles to Building a Beautiful World*, Chatto and Windus, London.

Easlea, Brian (1983). *Fathering the Unthinkable: Masculinity, Scientists and the Nuclear Arms Race*, Pluto Press, London.

Easlea, Brian (1984). 'Conflict, Science and the Garden of Eden', 2nd Keith Roby Memorial Lecture in Community Science, Murdoch University, W.A.

Eckersley, R. (1988a). 'Green Politics and the New Class. Selfishness or Virtue?', *Political Studies* (U.K.), Vol. 37, 1989, pp. 205–223.

Eckersley, R. (1988b). 'Divining Evolution. The Ecological Ethics of Murray Bookchin', *Environmental Ethics*, Vol. 11, No. 2, 1989, pp. 99–116.

Eckersley, R. (1989). 'The Paradox of Ecofeminism' in Dyer and Young (eds) *Ecopolitics IV Proceedings*, Adelaide, (1990).

Ehrlich, Paul (1962). *The Population Bomb*, Ballantine Books, New York.

Ehrlich, P., Ehrlich, A. and Holdren, J. (1973). *Human Ecology, Problems and Solutions*, W.H. Freeman, San Francisco.

Ehrlich, P., Ehrlich, A. and Holdren, J. (1977) *Ecoscience: Population, Resources Environment*, W.H. Freeman, San Francisco.

Ehrlich, P. and Holdren, J.P. (1988). *The Cassandra Conference: Resources and the Human Predicament*, Texas University Press, Austin.

Elliot, R. and Gare, A (eds) (1983) *Environmental Philosophy: A Collection of Readings*, University of Queensland Press, St Lucia, Queensland.

Ellul, Jacques (1964). *The Technological Society*, translated by John Wilkinson, Random House, New York.

Eysenck, H.J. (1971). *Race, Intelligence and Education*, Temple Smith, London.

Eysenck, H.J. (1973). *The Inequality of Man*, Temple Smith, London.

Faber, Daniel and O'Connor, J. (1989). *Struggles over Nature: The Environmental Crisis and the Crisis of Environmentalism in the U.S.*, Occasional Paper, Department of Sociology and Economics, University of California, Santa Cruz.

Fernbach, D. (1978). *The Alternative in Eastern Europe* (translation of Rudolph Bahro Die Allerndine 1977) N.L.B. Verso, London.

Fiji Government (1896). *Results of an Enquiry into the Causes of Native Depopulation*, Fiji National Archives.

Fisher, F.G. (1985). 'Overcoming Despair and the Alienation it Produces', *Australian Journal of Environmental Education*, 1:2, 16–20.

Fisk, E.K. (1974). *The Political Economy of Independent Fiji*, ANU Press, Canberra.

Fisher, F.G. and Hoverman, S. (1989) 'Environmental Science: Stirrings Toward a Science of Context' *International Journal of Environmental Education and Information*, Environmental Institute, University of Salford, vol. 8, no. I.

Flanagan, Sabina (1988). *Hildegard of Bingen: The Psychological and Social Uses of Prophecy*, Allen and Unwin, Sydney.

Fox, Matthew (1983). *Original Blessing: A Primer in Creation Spirituality*, Bear & Co., Santa Fe, NM.

Fox, Warwick (1984). 'Deep Ecology: A New Philosophy of Our Time?', *Ecologist*, 14:5–6, 194–200.

Fox, Warwick (1989). 'The Deep Ecology-Eco-Feminism Debate and its Parallels. *Environmental Ethics*, Vol. 11, No. 1, 5–25.

Fox, Warwick (1989). 'The Meanings of Deep Ecology', *Island Magazine*, no. 38, 32–6.

Fox, Warwick (1990). *Toward a Transpersonal Ecology*, Shambhala Publications, Boston, USA.

France, Peter (1969). *The Charter of the Land*, Oxford University Press.

Frank, Andre Gundar (1966). 'The Development of Underdevelopment', *Monthly Review*, September.

Frank, Andre Gundar (1967). *Capitalism and Underdevelopment in Latin America*, Monthly Review Press, New York.

Frankel, Boris (1989). 'Beyond Abstract Environmentalism', *Island Magazine*, no. 38.

Fromm, E. (1976). *To Have or To Be*, Abacus, London.

Galbraith, John K. (1978). *The New Industrial State*, 3rd edn, Houghton Miflin & Co, Boston.

Gilby, T. (ed.) (1963–75). *Summa Theologiae* by St Thomas

Aquinas, Blackfriars, London.

Glacken, C.J. (1970). *Traces on the Rhodian Shore: Nature and Culture in Western Thought from Ancient Times to the end of the 18th Century*, University of California Press, Berkeley.

Goldsmith, E. (1972). *A Blueprint for Survival*, The Ecologist, London.

Goldsmith, E. (1978). 'The Ecological Approach to Unemployment', *Ecologist Quarterly*, Spring, 32–75.

Goldsmith, E. (1985). 'Is Development the Solution or the Problem?' *Ecologist*, 15:5–6, 210–19.

Goldsmith, E. (1988a). *Battle for the Earth: Today's Key Environmental Issues*, Child and Associates, Brookvale, NSW.

Goldsmith, E. (1988b). *The Great U-Turn*, Green Books, Hartland, Devon.

Goldsmith, E. and Hildyard, N. (1988). *The Social and Environmental Effects of Large Dams* (3 vols), Ecosystems Ltd, Camelford, Cornwall.

Goodland, J.A., Watson, C. and Ledec, G. (1984). *Environmental Management in Tropical Agriculture*, Westnew Press, Boulder, Colo.

Gorz, A. (1983) *Ecology as Politics*, translated by P. Vigderman and J. Cloud, Pluto Press, London.

Gorz, A. (1985). *Paths to Paradise: On the liberation from Work*, translated by Malcolm Imrie, Pluto Press, London.

Green Party (1986). *Manifesto for a Sustainable Society*, Green Party, London.

Greer, Germaine (1985). *Sex and Destiny: The Politics of Human Fertility*, Picador, London.

Griffin, Susan (1978). *Woman and Nature*, Harper and Row, New York.

Hallen, Patsy (1987). 'Making Peace with Nature: Why Ecology needs Feminism', *The Trumpeter*, 4:3.

Hardin, G. (1968). 'The Tragedy of the Commons', *Science*, vol. 162, pp. 1243–8.

Hardin, Garrett (1977). *The Limits to Altruism*, Indiana University Press, Indianapolis.

Hammond, J.L. and B. (1919). *The Skilled Labourer 1760–1832*, Longman, London.

Hardy, G.H. (1940). *A Mathematician's Apology*, Cambridge University Press.

Harrison, P. (1980). *Inside the Third World: An Anatomy of Poverty*, Harvester Press, Brighton.

Hay, Peter, Eckersley, Robyn and Holloway, Geoff (eds) (1989) *Environmental Politics in Australia and New Zealand*, Centre for Environmental Studies, University of Tasmania.

Heasley, M. (1982). 'The Life and Times of Cakobau', Ph.D. thesis, University of Otago, New Zealand.

Herber, Lewis (pseud.) (1962). *Our Synthetic Environment*, Arnold Knopf, New York.

Hirsch, F. (1977). *Social Limits to Growth*, Routledge and Kegan Paul, London.

Home, R.W. (ed.) (1983) *Science Under Scrutiny: The Place of History and Philosophy of Science*, D. Reidel Publishing Co, Dortrecht, Boston, Lancaster.

Hopa, Npare (1990). 'Papatuanuku: Spaceship Earth' in Dyer and Young (eds), *New Directions Ecopolitics IV Proceedings*, University of Adelaide.

Howe, Kerry (1974). 'Firearms and Indigenous Warfare: A Case Study' *Journal of Pacific History*, vol. 9, pp. 21–39, Oxford University Press.

Howe, Kerry (1977). *The Loyalty Islands: A History of Culture Contacts*, ANU Press, Canberra.

Hughes, D.J. (1975). *Ecology in Ancient Civilisations*, Albuquerque, NM.

Hughes, D.J. and Thurgood, J.V. (1982). 'Deforestation in Ancient Greece and Rome: A Cause of Collapse', *Ecologist*, 12:5.

Hutchinson, R. (1986). *A Sustainable Economy: The Green Commitment*, Liberal Party, London.

Hutton, Drew (ed.) (1987). *Green Politics in Australia*, Angus and Robertson, Sydney.

Illich, Ivan (1973). *Tools for Conviviality*, Harper and Row, New York.

Jarvie, I.C. (1964). *The Revolution in Anthropology*, Routledge and Kegan Paul, London.

Jensen, A.R. (1969) 'How much can we boost IQ and Scholastic Achievement?, *Harvard Educational Review*, Winter 1969.

Jensen, R. (1972). *Genetics and Education*, Methuen, London.

Kahn, H. and Bruce-Biggs, R. (1972). *Things to Come: Thinking About the Seventies and Eighties*, Macmillan, New York.

Kahn, H. and Weiner, A.J. (1967). *The Year 2000: A framework for speculation on the next 33 years*, Macmillan, New York.

Keynes, J.M. (1930). 'The Economic Possibilities for Our Grandchildren' in *Collected Writings*, Macmillan, London (1972), vol. IX, 321–32.

Knight, D. (1987). *The Age of Science*, Oxford University Press, Oxford.

Komarov, Boris (1980). *The Destruction of Nature in the Soviet Union*, White Plains, New York.

Koestler, A. (1968). *The Sleepwalkers*, Macmillan, New York.

Kropotkin, P. (1968) (1899). *Memoirs of a Revolutionist*, Horizon

Press, New York.

Kuhn, Thomas (1962). *The Structure of Scientific Revolutions*, University of Chicago Press, Chicago and London.

Lall, S. and Stewart, F. (1986). *Theory and Reality and Development Essays in Honour of Paul Streeten*, Macmillan, Basingstoke.

Larner, Christine (1981). *Enemies of God*, Chatto and Windus, London.

Lee, D.C. (1980). 'On the Marxian View of the Relationship between Man and Nature', *Environmental Ethics*, no. 2, 3–16.

Leopold, A. (1949). *A Sand County Almanac*, Oxford University Press, New York.

Levins, R. and Lewontin, R. (1986). *The Dialectical Biologist*, Harvard University Press, Cambridge, Mass.

Lovejoy, A.O. (1936). *The Great Chain of Being*, Harvard University press, Cambridge, Mass.

Lovelock, J.E. (1979). *Gaia: A New Look at Life on Earth*, Oxford University Press.

Lovelock, J.E. (1988). *The Ages of Gaia: A Biography of Our Living Earth*, Norton, New York.

Lovins, A. (1979). *Soft Energy Paths*, Harper and Row, New York.

Lundberg, G.A., Bain, R., Anderson, N. (1929). *Trends in American Sociology*, Harper, New York.

Lutzenberger, J.A. (1982). 'The Systematic Demolition of the Tropical Rainforest of the Amazon' *Ecologist*, 12:6, 248–52.

McArthur, Norma (1968). *The Island Populations of the Pacific*, ANU Press, Canberra.

McCormick, E.H. (ed.) (1963). *New Zealand, or Recollections of it by Edward Markham*. N.Z. Department of Internal Affairs, Wellington.

McCulloch, A. and Abrams, P. (1976). *Communes, Sociology and Society*, Cambridge University Press.

McDonagh, S. (1986). *To Care for the Earth*, Geoffrey Chapman, London.

McPhee, John (1971). *Encounters with the Arch Druid: Narratives about a Conservationist and three of his Natural Enemies*. Farrar, Strauss and Giroux, New York.

McRobie, G. (1981). *Small is Possible*, Jonathan Cape, London.

Maddox, John (1972). *The Domesday Syndrome*, Maddox Editorial Ltd., London.

Mair, Lucy (1984). *Anthropology and Development*, Macmillan, London.

Mansell, Michael (1987). 'Conservationists: Trespassers on Aboriginal land' in *Ecopolitics II Proceedings*, University of Tasmania.

Marcuse, H. (1969(. *Eros and Civilisation: A Philosophical Inquiry into Freud*. Sphere Books, London.

Marcuse, Herbert (1964). *One Dimensional Man: Studies in the Ideology of Advanced Industrial Society*, Routledge and Kegan Paul, London.

Martin, Brian (1980). *Changing the Cogs*, privately printed, Canberra.

Martin, Brian (1981a). 'The Scientific Straightjacket: The Power Structure of Science and the Suppression of Environmental Scholarship', *Ecologist*, 11, 33–44.

Martin, Brian (1981b). *The Bias of Science*, privately printed, Canberra.

Martin, Brian (1982). 'The Naked Experts', *Ecologist*, 12:4.

Martin, Malachi (1975). 'Ravings of a Fatigued, Drunken, Young Ex-scientist' in Clarke (1975), 119 ff.

Marwick, Max (ed.) (1970). *Witchcraft and Sorcery*, Penguin Books, London.

Marx, K. and Engels, F. (1969). *Selected works* (3 vols), Progress Publishers, Moscow.

Maude, H.E. (1968). *Of Islands and Men*, Oxford University Press.

Mead, Margaret (1939). *From the South Seas: Studies of Adolescence and Sex in Primitive Societies*, William Morrow, New York.

Merchant, Carolyn (1980). *The Death of Nature: Women, Ecology and the Scientific Revolution*, Harper and Row, San Francisco.

Merton, R.K. (ed.) (1973). *The Sociology of Science*, University of Chicago Press.

Mills, J.S. (1891). *Principles of Political Economy*, George Routledge, London.

Mills, Stephanie (1981). 'Planetary Passions: A Reverent Anarchy' *Co-evolution Quarterly*, no. 32, Winter (Bio-regions Special Issue).

Moorehead, Alan (1966). *The Fatal Impact: An Account of the Invasion of the South Pacific 1767–1840*, Hamish Hamilton, London.

Morgan, Robin and Whitaker, Brian (1986). *Rainbow Warrior*, Sunday Times Insight Team, London.

Morton, A.L. (ed.) (1973). *The Political Writings of William Morris*, Lawrence and Wishart, London.

Mundey, Jack (1987). 'From Red to Green: Citizen Worker Alliance' in Hutton, D. (ed.) *Green Politics in Australia*, pp. 105–23.

Mundey, Jack (1990). 'From Grey to Green' in *Ecopolitics IV Proceedings*, Dyer and Young (1990) (eds) *New Directions*, Adelaide.

Munro-Clark, Margaret (1986). *Communes in Rural Australia: The Movement Since 1970*, Hale and Iremonger, Sydney.

Murdoch, W.W. (1980). *The Poverty of Nations: The Political Economy of Hunger and Population*. John Hopkins Press, Baltimore, Md.

Murray, Margaret A. (1921). *The Witch Cult in Western Europe*, Clarendon Press, Oxford.

Myrdal, Gunnar (1957). *Economic Theory and Underdeveloped Regions*, Duckworth, London.

Naess, Arne (1973). 'The Shallow and the Deep, Long-range Ecology Movement, A Summary', *Inquiry*, 16, 95–100.

Naess, Arne (1985). 'Identification as a source of Deep Ecological Attitudes' in Tobias (1985).

Naess, Arne (1986). 'The Deep Ecological Movement: Some Philosophical Aspects', *Philosophical Inquiry*, 8, 10–31.

Newbury, C.W. (ed.) (1961). *The History of the Tahitian Mission 1799–1830* by John Davies, Cambridge University Press. Hakluyt Society, London

O'Connor, James (1987). *The Meaning of Crisis: A Theoretical Introduction*, Basil Blackwell, Oxford.

Ophuls, W. (1977). *Ecology and the Politics of Scarcity: Prologue to a Political Theory of the Steady State*, W.H. Freeman, San Francisco.

Paddock, W. and Paddock, P. (1967). *Famine 1975! Who Will Survive?*, Little, Brown, Boston.

Parkin, Frank (1968). *Middle Class Radicalism: The Social Bases of the British Campaign for Nuclear Disarmament*, Manchester University Press.

Passmore, John (1974). *Man's Responsibility for Nature*, Duckworth, London.

Patrick, C. (1988). 'Australian Environmental Studies, the Cinderella Subject', *Australian Studies*, 9, 25–32.

Patterson, T. (1984). *Conservation and the Conservatives*, Conservative Party, London.

Payne, F. (1947). 'The Plough in Ancient-Britain', *Archaeological Journal*, Royal Archaeological Institute, London.

Peattie, Lisa (1986). 'The Defence of Daily Life' International Union of Anthropological and Ethnographical Sciences Commission on the Study of Peace, *Newsletter*, 4.

Pepper, D. (1984). *The Roots of Modern Environmentalism*, Croom Helm, London.

Pepper, D. (1986). 'Radical Environmentalism and the Labour Movement', in Joe Weston (ed.), *Red and Green the New Politics of the Environment*, Pluto Press, London.

Phelps, S. (1976). 'A Study of Valuables', Ph.D. thesis Cambridge University.

Plumwood, V. and Routley, R. (1982). 'World Rainforest

Destruction, the Social Factors', *Ecologist*, 12:1, 4–23.

Pollard, Nigel (1984). 'The Israelites and their Environment', *Ecologist*, 14:3, 125–33.

Porritt, Jonathon (1984). *Seeing Green: The Politics of Ecology Explained*, Basil Blackwell, Oxford.

Porritt, Jonathon (1987). *Friends of the Earth Handbook*, Optima, London.

Porritt, Jonathon (1988). 'Let the Green Spirit Live' (Schumacher Lecture, 1987), *Resurgence*, no. 127.

Porritt, J. and Winner, D. (1988). *The Coming of the Greens*, Fontana, London.

Prigogine, Ilya and Stengers, Isobelle (1984). *Order Out of Chaos, Man's New Dialogue with Nature*, Random House, Boulder, Colorado.

Rappaport, R. (1967). *Pigs for the Ancestors: Ritual in the Ecology of a new Guinea People*, Yale University Press, New Haven, Conn.

Rappaport, R. (1979). *Ecology, Meaning and Religion*, North Atlantic Books, Richmond, Calif.

Ravuvu, Asesela (1984). *The Fijian Ethos*, University, of the South Pacific, Suva.

Ravuvu, Asesela (1984). *The Fijian Way of Life*, University of the South Pacific, Suva.

Reich, Charles (1970). *The Greening of America*, Random House, New York.

Rika, Nacalieni (1974). 'Is Kinship Costly', *Pacific Perspective*, Suva.

Rodney, W. (1981). *How Europe Underdeveloped Africa*, Howard University Press, Washington D.C.

Rose, Hilary and Steven (1969). *Science and Society*, Allen Lane, London.

Rose, Steven and Hilary (1976). *The Radicalism of Science: Ideology of/in the Natural Sciences*, Macmillan, London.

Rostow, W.W. (1960). *The Stages of Economic Growth*, Cambridge University Press.

Roszak, T. (1973). *Where the Wasteland Ends: Politics and Transendence in Post Industrial Society*, Faber and Faber, London.

Rothschild, J. (ed.) (ex-ref.). *Machina Ex Dea: Feminist Perspectives on Technology*, Pergamon Press, New York.

Rothschild, N.M.V. (1971). *The Organisation of Government R and D*, Cmnd 4814, HMSO, London.

Ruether, Rosemary (1975). *New Woman, New Earth*, Seabury Press, New York.

Rust, Eric (1953). *Nature and Man in Biblical Thought*, Lutterworth Press, London.

St Johnstone, T.R. (1910). M.S. Fiji National Archives, Suva. Colonial Secretary's Office Inwards Correspondence. Comments on Vital Statistics for Lau 9287/10.

St Johnstone, T.R. (1913). M.S. Fiji National Archives, Suva. District Commissioner's Annual Report for Lau, 1912. 22 January 1913. Colonial Secretary's Office, Inwards Correspondence 688/1913.

St Johnstone, T.R. (1914). M.S. Fiji National Archives, Suva. Colonial Secretary's Office Inwards Correspondence. Annual Report for Lau, 1913. January 1914. 1349/1914.

Sagoff, Mark (1988). *The Economy of the Earth: Philosophy, Law and the Environment*, Cambridge University Press, Cambridge and New York.

Sahlins, M. (1985). *Islands of History*, Chicago University Press, Chicago.

Sale, Kirkpatrick (1984). 'Bio-regionalism, A New Way to Treat the Land', *Ecologist*, 14:4.

Salleh, Ariel (1984). 'Deeper than Deep Ecology: the Eco-feminist Connection', *Environmental Ethics*, no. 6.

Salzman, Lorna (1983). 'Ecology and Politics in the United States', *Ecologist*, 13: 40–2.

Salzman, Lorna (1990). 'Politics as if Evolution Mattered: Some Thoughts on Deep Ecology and Social Ecology' in Dyer and Young (eds) *New Directions*: Ecopolitics IV proceedings, University of Adelaide.

Sandbach, Francis (1980). *Environment: Ideology and Policy*, Basil Blackwell, Oxford.

Scherer, D. and Attig, T. (eds) (1983). *Ethics and the Environment*, Prentice-Hall, NJ.

Schnaiberg, Alan (1980). *The Environment: From Surplus to Scarcity*, Oxford University Press.

Schneider, B. (1986). *The Barefoot Revolution*, Intermediate Technology Publications, London.

Schumacher, E.F. (1955). *Economics in a Buddhist Country*, Rangoon.

Schumacher, E.F. (1974). *Small is Beautiful*, Abacus Books, London.

Schumacher, E.F. (1977). *A Guide for the Perplexed*, Jonathan Cape, London.

Schumacher, E.F. (1979). *Good Work*, Jonathan Cape, London.

Scitovsky, Tabor (1976). *The Joyless Economy: An Inquiry into Human Satisfaction and Consumer Dissatisfaction*, Oxford University Press.

Seed, John (ed.) (1988). *Thinking Like a Mountain*, New Society Publishers, Philadelphia.

Sen, A. (1986). 'Adam Smith's Prudence' in Lall, Sanjaya and Steward, Frances, *Theory and Reality in Development*, Macmillan, London.

Sessions, G. (1974). 'Anthropocentrism and the Environmental Crisis', *Humbolt Journal of Social Relations*, 2, 71–81.

Sessions, G. (1977). 'Spinoza and Jeffers on Man in Nature', *Inquiry*, no. 20, 481–528.

Sherrard, Philip (1987). *The Rape of Man and Nature: An Enquiry into the Origins and Consequences of Modern Science*, Golgonooza Press.

Shineberg, D. (1967). *They Came for Sandalwood*, Melbourne University Press.

Shockley, W. (1971). 'Negro IQ deficit: Failure of a "malicious co-incidence warrants new research proposals"', *Review of Educational Research*, 41, pp. 227–48.

Sigurdson, Jon (1977). *Rural Industrialisation in China*, Harvard University Press, Cambridge, Mass.

Simon, Julian L. and Kahn, H. (eds) (1984). *The Resourceful Earth: A response to Global 2000*, Basil Blackwell, Oxford and Oxford University Press.

Snow, C.P. (1962). *The Two Cultures and the Scientific Revolution*, Cambridge University Press.

Snow, C.P. (1965). *The Two Cultures: A Second Look*, Cambridge University Press.

Spretnak, Charlene and Capra, Fritjof (1985). *Green Politics: The Global Promise*, Paladin Books, London.

Stephen, Joseph (1986). 'Trade Unions: Green as a Class Issue', *New Ground*, no. 10, Summer.

Stickland, David (1986). Organic Farmers and Growers Information Leaflet, Needham Market, Suffolk.

Streeten, Paul (1981). *Development Perspectives*, Macmillan, London.

Streeten, Paul and Lall, S. (1976). *Foreign Investment, Trans-nationals and Developing Countries*, Macmillan, London.

Stretton, Hugh (1969). *The Political Sciences: General Principles of Selection in Social Science and History*, Routledge and Kegan Paul.

Stretton, Hugh (1970). *Ideas for Australian Cities*, Hugh Stretton, North Adelaide, for the benefit of the Urban Social Services of the Brotherhood of St Laurence, Melbourne.

Stretton, Hugh (1976). *Capitalism, Socialism and the Environment*, Cambridge University Press.

Stretton, Hugh (1978). *Urban Planning in Rich and Poor Countries*, Oxford University Press, Oxford and New York.

Stretton, Hugh (1987). *Political Essays*, Georgian House, Melbourne.

Sullivan, Andrew (1985). *Greening the Tories: New Policies for the Environment*, Conservative Policy Centre, London.

Sutherland, William (1918). Fiji National Archives, Suva. Colonial Secretary's Office Inwards Correspondence 569/1918. Annual Report for Lau, 1917.

Swift, J. (1726). *Travels into several Remote Nations of the World, in four parts, by Lemuel Gulliver, First a Surgeon, and then a Captain of Several Ships*, Benj Motte, London.

Sylvan, R. (1985). 'A critique of Deep Ecology', *Radical Philosophy*, no. 40, 2–12; no. 41, 10–22..

Tawney, R.H. (1938). *Religion and the Rise of Capitalism*, Pelican Books, London.

Thomas, Keith (1971). *Religion and the Decline of Magic*, Weidenfeld and Nicolson, London.

Thomas, Keith (1983). *Man and the Natural World: Changing Attitudes in England 1500–1800*, Allen Lane, London.

Thomis, M.I. (1974). *The Town Labourer and the Industrial Revolution*, B.T. Batsford, London and Sydney.

Thompson, E.P. (1980). *The Making of the English Working Class*, Penguin, London.

Tobias, Michael (ed.) (1985). *Deep Ecology*, Avant Books, San Diego.

Tokar, Brian (1987). *The Green Alternative*, R. and E. Miles, New York.

Trevor-Roper, H.R. (1972). *Religion, the Reformation and Social Change, and other essays*, Macmillan, London.

Trussell, Dennis (1982). 'History of an Antipodean Garden', *Ecologist*, 12:1.

Tucker, W. (1980). 'Environmentalism, the Newest Toryism', *Policy Review*, 14, 141–52.

Tunlawiroon, N. (1985). 'The Environmental Impact of Industrialisation in Thailand', *Ecologist*, 15:4, 161–4.

Vogt, William (1949). *Road to Survival*, Victor Gollancz, London.

Waites, Bryan (1984). 'In Search of Britain's Regions', *Ecologist*, 14:4, 161–6.

Weston, Joe (ed.) (1986). *Red and Green: The New Politics of the Environment*, Pluto Press, London.

Whitehead, Alfred N. (1925). *Science and the Modern World*, MacMillan, New York.

White, Lynn (1962). *Medieval Technology and Social Change*, Clarendon Press, Oxford.

White, Lynn (1967). 'The Historical Roots of Our Environmental Crisis', *Science*, March, 1204 ff.

Wild Goose Publications (1985). *What is the Iona Community*, Iona.

Wilson, Charlotte (1896). *What is Socialism?*, Fabian Tract, no. 4.

Winstanley, Gerrard (1649). 'The True Leveller's Standard Advanced' in *The Law of Freedom and other Writings*, edited with an introduction by Christopher Hill, Cambridge University Press, Cambridge and New York 1983.

Wiseman, J. (1984) 'Red or Green? The German Ecological Movement', *Area* (Melbourne), no. 68.

Wood, Barbara (1984). *Alias Papa: A Life of Fritz Schumacher*, Jonathan Cape, London.

Woodcock, G. and Avakumovic (1950). *The Anarchist Prince: A Biographical Study of Peter Kropotkin*, T.V. Boardman, London, New York.

Worster, Donald (1987). *Nature's Economy: The Roots of Ecology*, 2nd edn, Cambridge University Press.

Wright, H.M. (1959). *New Zealand 1769–1840: Early Years of Western Contact*, Harvard University Press, Cambridge, Mass.

Yett, J. (1984) 'Women and their Environment', *Environmental Review*, 8, 86–94.

Young, John M.R. (1987). 'From Productivity to Creativity: Technological Predestination and Political Free Will' in Fischer, F. (ed.), *Sustaining Gaia: Contributions to Another World View*, Monash University.

Young, John, Dyer, K.F. and Taylor, Sandra (1989) 'The Politics of Environmental Studies' in Dixon, Jennifer, Gunn, A. and Erickson, N.J. Ecopolitics III Proceedings Waikato University, New Zealand, September 1988.

Young, Susan (1989). 'Policy, Practice and the Private Sector in China', *Australian Journal of Chinese Affairs*, no. 21, pp. 57–80.

Zablocki, B.D. (1971). *The Joyful Community*: An account of the Brudenhof, a communal movement now in its third generation, Penguin Books, Baltimore.

Index